Multiscale, Multiphysics Modelling of Coastal Ocean Processes: Paradigms and Approaches

Multiscale, Multiphysics Modelling of Coastal Ocean Processes: Paradigms and Approaches

Editors

Hansong Tang
C. Reid Nichols
Donald T. Resio
Don Wright

MDPI • Basel • Beijing • Wuhan • Barcelona • Belgrade • Manchester • Tokyo • Cluj • Tianjin

Editors

Hansong Tang
Department of Civil
Engineering, The City College of
New York, 160 Convent Avenue,
New York
USA

C. Reid Nichols
Marine Information Resources
Corporation, 12337 Pans Spring
Court, Ellicott City
USA

Donald T. Resio
College of Computing,
Engineering & Construction,
University of North Florida,
1 UNF Drive, Jacksonville
USA

Don Wright
Southeastern Universities
Research Association, 1201 New
York Ave. NW, Suite 430,
Washington
USA

Editorial Office
MDPI
St. Alban-Anlage 66
4052 Basel, Switzerland

This is a reprint of articles from the Special Issue published online in the open access journal *Journal of Marine Science and Engineering* (ISSN 2077-1312) (available at: https://www.mdpi.com/journal/jmse/special_issues/mul_model_coastal).

For citation purposes, cite each article independently as indicated on the article page online and as indicated below:

LastName, A.A.; LastName, B.B.; LastName, C.C. Article Title. *Journal Name* **Year**, *Volume Number*, Page Range.

ISBN 978-3-0365-2810-6 (Hbk)
ISBN 978-3-0365-2811-3 (PDF)

Contents

Preface to "Multiscale, Multiphysics Modelling of Coastal Ocean Processes: Paradigms and Approaches"

Climate change and increasing human activities have resulted in many ocean flow problems, which involve distinct physical processes across varying temporal and spatial scales and present grand challenges to our modeling capabilities. For instance, the 2010 BP oil spill in the Gulf of Mexico started as a fully three-dimensional, high-speed jet flow covering an area of 10 m. Later, it evolved into drifting patches of oil film on the water surface, expanding to hundreds of kilometers of coverage. Although both are fluid flows, the jet and patches comprise multiscale, multiphysics phenomena; they exhibit distinct physical behaviors at vastly different scales, and different sets of governing equations better describe them. More example problems include the interaction of land runoff and coastal water, compound flooding, and tsunami propagation and its impact on coastal structures. Currently, the state-of-the-art simulations for these phenomena (e.g., spill jet and oil patches), are made using wholly disjoint computer models (e.g., from the engineering and ocean science communities). However, these approaches encounter difficulties in appropriately handling such multiscale, multiphysics problems to meet the needs of scientific research and engineering practice. It has become critical to break through barriers and to integrally and simultaneously simulate these flow phenomena and their interactions.

Multiscale simulations of coastal ocean flows trace back to early mariners interested in determining optimal shipping routes owing to the impacts of permanent currents such as the Gulf Stream and changing sea states. Now, such simulations have become common practice within the ocean science community. However, this is not yet the case for multiphysics simulations, which in general cannot be realized merely by local mesh refinement but needs to integrate different models. To attract the attention of communities and to promote further development, this Special Issue of the *Journal of Marine Science and Engineering* entitled "Multiscale, Multiphysics Modelling of Coastal Ocean Processes: Paradigms and Approaches" is a collection of papers on the simulation of multiscale, multiphysics coastal ocean flows. This collection provides perspectives on the status of such simulations, discussions related to current issues, and research ideas to further understand this field of science. This Special Issue covers coastal flooding, model assessment, effects of scales and wind fields, model coupling, parallel computation, and computational methods.

We thank all authors who kindly contributed their papers to this issue and the in-house editors from the *Journal of Marine Science and Engineering* for their kind help and co-operation. We are also indebted to the MDPI staff for their assistance in preparing and publishing this issue.

Hansong Tang, C. Reid Nichols, Donald T. Resio, Don Wright
Editors

Article

Contrasting Hydrodynamic Responses to Atmospheric Systems with Different Scales: Impact of Cold Fronts vs. That of a Hurricane

Wei Huang [1,*] and Chunyan Li [2]

[1] Virginia Institute of Marine Science, College of William and Mary, Gloucester Point, VA 23062, USA

[2] Department of Oceanography and Coastal Sciences, School of Coast and Environment, Louisiana State University, Baton Rouge, LA 70803, USA; cli@lsu.edu

* Correspondence: whuang@vims.edu

Received: 1 November 2020; Accepted: 25 November 2020; Published: 2 December 2020

Abstract: In this paper, subtidal responses of Barataria Bay to an atmospheric cold front in 2014 and Hurricane Barry of 2019 are studied. The cold fronts had shorter influencing periods (1 to 3 days), while Hurricane Barry had a much longer influencing period (about 1 week). Wind direction usually changes from southern quadrants to northern quadrants before and after a cold front's passage. For a hurricane making its landfall at the norther Gulf of Mexico coast, wind variation is dependent on the location relative to the location of landfall. Consequently, water level usually reaches a trough after the maximum cold front wind usually; while after the maximum wind during a hurricane, water level mostly has a surge, especially on the right-hand side of the hurricane. Water level variation induced by Hurricane Barry is about 3 times of that induced by a cold front event. Water volume flux also shows differences under these two weather types: the volume transport during Hurricane Barry was 4 times of that during a cold front. On the other hand, cold front events are much more frequent (30–40 times a year), and they lead to more frequent exchange between Barataria Bay and the coastal ocean.

Keywords: cold front; Hurricane Barry; numerical simulation; subtidal hydrodynamics; multi-inlet; volume flux

---

1. Introduction

The coast of Louisiana in the northern Gulf of Mexico (NGOM) is characterized by semi-enclosed bays with exchange flows of water through multiple inlets, such as Lake Pontchartrain, Calcasieu Lake, Vermilion Bay, and Barataria Bay. They have limited connections with the coastal ocean except through narrow inlets. These are, however, different from inland freshwater lakes or general coastal plain estuaries connected to the coastal ocean through multiple inlets. The NGOM has several major environmental processes that are determined by hydrodynamics, particularly those related to the exchange of water and sediment between estuaries and shelf water, e.g., the significant land loss around lower Mississippi River basin associated with processes that cause erosion and sediment transport [1–6]. Along the NGOM coast, the most regular hydrodynamic motions are the relatively weak tides, which are mainly diurnal oscillations with a maximum tidal range of about 0.6 m [7,8]. Because of weak tides, the effect of weather [9–14] becomes more prominent in moving the sediment through inundation and erosion [15]. As a result, the less predictable weather-induced bay oscillations may cause more significant flood and drain of the micro-tidal system [16–20]. Synoptic weather systems and hurricanes can produce responses in these water bodies affecting the water exchange, which is important to the ecosystem [21–23]. However, there is a lack of in-depth analysis of weather conditions characterizing different weather patterns.

Among the weather influences, hurricane impact can be most dramatic. Because of the low gradient of land and relatively shallow and broad shelf along NGOM, hurricanes can cause significant damages to the micro-tidal coast zone [24–28]. For instance, severe storm surge caused damages in 2005 by Hurricanes Katrina and Rita [29–33], in 2008 by Hurricanes Gustav and Ike [34–37], and in 2017 by Hurricanes Harvey and Irma [38]. Compared with hurricanes, cold front-associated winds can be more frequent, and have a more accumulative effect in driving the hydrodynamics. The length of influence of a cold front can reach ~2000–3000 km, much larger [39] than the region of strong wind within a typical hurricane. Previous studies have covered various aspects of weather impact to coastal ocean. For example, Keen [40] used numerical models to predict the waves and currents under cold front passage over Mississippi bight. Keen and Stavn [41] later used observations and numerical models with interaction of atmospheric forcing and hydrodynamics to investigate the optical environment at Santa Rosa Island, Florida during two cold front passages. Water exchange and circulations in the bays and estuaries under meso-scale weather systems like winter storms and cold fronts can be related not only to the circulations in the coastal regions but also to the sediment transport [42] and ecosystems. Sediment transport and distributions on the shallow shelf and in the estuaries of Gulf of Mexico under the influence of cold front passages in winter time are investigated by, e.g., Perez et al. [43] and Kineke et al. [44]. Siadatmousavi et al. [45] studied the wave energy during a cold front and skill assessed a phase-averaged spectral wave model.

There have been many studies on subtidal flow in estuaries [46–48]. In these studies, the wind effects are often discussed as a time series forcing without examining the spatial structure of the weather systems. The subtidal energy in the estuarine circulations caused by cold fronts may be comparable if not larger than that of tides in the area [49]. A recent study [50] investigated the weather-induced exchange flows through multiple inlets of the Barataria Bay in a few months period in 2013, 2014, and 2015 with 51 atmospheric cold fronts passing the Louisiana coast. These events are apparently very common: an analysis [51] covering a period of 40 years identified more than 1600 atmospheric frontal events, with an average of ~41.2 ± 4.7 per year excluding the months between May and August for much weaker activities of this kind. However, no quantitative comparisons are made between the hydrodynamic responses induced by cold fronts and that from hurricanes. Therefore, it is of interest for a comparison between the hydrodynamic responses to these two different weather systems with different scales.

This work will use a calibrated three-dimensional finite volume community model (FVCOM) to simulate water level and flows in Barataria Bay under multi-scale weather systems including cold front and hurricane events. The goals are to (1) compare the hydrodynamic responses to different weather systems (cold front and hurricane), (2) examine water exchange between Barataria Bay and coastal ocean through multiple inlets under the different weather systems, and (3) assess the quasi-steady state balance under different weather systems.

2. Study Site and Data

Barataria Bay (Figure 1) is a shallow estuary in southeast Louisiana and south of the City of New Orleans. It is bounded by several barrier islands and irregular-shaped wetlands with multiple tidal inlets connecting to the open ocean. The main axis from north to south and from east to west is about 30–40 km. The tidal inlets include Barataria Pass with a width of ~800 m and a maximum depth of 20 m at the mouth, Caminada Pass with ~800 m width, 9 m depth, and a 90 degree turn in channel orientation near the mouth, and the 15 m deep Pass Abel with a width of about 1.9 km. Freshwater is mainly from the manmade Davis Pond Diversion facility with a capacity of about 250 m^3/s of flux. Water inside the Barataria Bay is very shallow (average depth of 2 m). Erosions in the bay appear to be significant, e.g., there is a 50 m hole [50] northwest of the Barataria Pass, which is the deepest point among all Louisiana lagoons, bays, and estuaries, revealing the significant contribution of non-tidal forcing to the micro-tidal system.

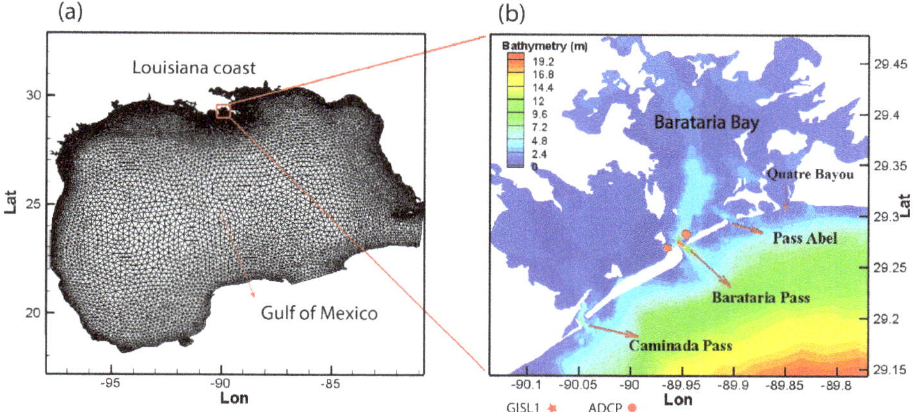

Figure 1. Study site (**b**) and model grid (**a**) for FVCOM simulation. Star represents the location of wind observations (NDBC station GISL1); red dot represents the location of ADCP deployment for water level and velocity observations.

Observational data of water level and velocity were obtained from 5 Sontek Argonaut DP SL 500-KHz horizontal acoustic Doppler current profilers (ADCPs, manufacture: SonTek/Xylem Inc., San Diego, USA). Information about the measurements can be found in Li et al. [50]. Wind data are from the National Ocean Service station at Grand Isle (29.265° N, 89.958° W, Figure 1). The atmospheric forcing for the upper boundary of the hydrodynamic numerical model was obtained from the global Climate Forecast System Reanalysis (CFSR) data (https://climatedataguide.ucar.edu/climate-data/climate-forecast-system-reanalysis-cfsr).

3. Model Setup and Validation

3.1. Model Description

A finite volume community ocean model (FVCOM) was applied in this study. FVCOM is widely used for investigating coastal ocean hydrodynamics with complicated topography [52]. The governing equations are [52]:

$$\frac{\partial u}{\partial t} + u\frac{\partial u}{\partial x} + v\frac{\partial u}{\partial y} + w\frac{\partial u}{\partial z} - fv = -\frac{1}{\rho_0}\frac{\partial P}{\partial x} + \frac{\partial}{\partial z}\left(K_m\frac{\partial u}{\partial z}\right) + F_u,\tag{1}$$

$$\frac{\partial v}{\partial t} + u\frac{\partial v}{\partial x} + v\frac{\partial v}{\partial y} + w\frac{\partial v}{\partial z} + fu = -\frac{1}{\rho_0}\frac{\partial P}{\partial y} + \frac{\partial}{\partial z}\left(K_m\frac{\partial v}{\partial z}\right) + F_v,\tag{2}$$

$$\frac{\partial w}{\partial t} + u\frac{\partial w}{\partial x} + v\frac{\partial w}{\partial y} + w\frac{\partial w}{\partial z} = -\frac{1}{\rho_0}\frac{\partial p}{\partial z} + \frac{\partial}{\partial z}\left(K_m\frac{\partial w}{\partial z}\right) + F_w - g,\tag{3}$$

$$\frac{\partial u}{\partial x} + \frac{\partial v}{\partial y} + \frac{\partial w}{\partial z} = 0,\tag{4}$$

where x, y, z are the three axes in the east, north, and vertical directions, respectively; u, v, w are the x, y, z velocities, respectively, ρ_0 is the density; P is the total pressure of air and water; f is the Coriolis parameter; g is the gravitational acceleration; K_m is the vertical eddy diffusion coefficient, determined by the Mellor and Yamada [53] level-2.5 (MY-2.5) turbulent closure scheme; F_w is the diffusion term of the vertical momentum; and F_u, F_v are the diffusion terms for the horizontal momentums.

The surface and bottom boundary conditions are:

$$K_m\left(\frac{\partial u}{\partial z}, \frac{\partial v}{\partial z}\right) = \frac{1}{\rho_0}\left(\tau_{sx}, \tau_{sy}\right), \ w = \frac{\partial \zeta}{\partial t} + \frac{\partial \zeta}{\partial x} + \frac{\partial \zeta}{\partial y}, \ at \ z = \zeta(x,y,t), \tag{5}$$

$$K_m\left(\frac{\partial u}{\partial z}, \frac{\partial v}{\partial z}\right) = \frac{1}{\rho_0}\left(\tau_{sx}, \tau_{sy}\right), \ w = -u\frac{\partial H}{\partial x} - v\frac{\partial H}{\partial y}, \ at \ z = -H(x,y), \tag{6}$$

where $\left(\tau_{sx}, \tau_{sy}\right)$ and $\left(\tau_{bx}, \tau_{by}\right)$ are the surface wind stress and bottom stress vectors, respectively. H is the water depth and ζ is the surface elevation. $\left(\tau_{sx}, \tau_{sy}\right)$ is calculated by $C_d\rho_a|U_{10}|U_{10}$, where U_{10} is the wind at 10 m height, ρ_a is the air density (1.29 kg/m^3), and C_d is the surface wind drag coefficient and is calculated by the following equations:

$$C_d = \begin{cases} (0.49 + 0.065 \times 11.0) \times 10^{-3}, \ U_{10} < 11.0 \ m/s \\ (0.49 + 0.065 \times |U_{10}|) \times 10^{-3}, \ 11.0\frac{m}{s} \le U_{10} \le 25.0 \ m/s \ , \\ (0.49 + 0.065 \times 25.0) \times 10^{-3}, \ U_{10} > 25.0 \ m/s \end{cases} \tag{7}$$

where $\left(\tau_{bx}, \tau_{by}\right)$ is the bottom stress calculated by $C_d \sqrt{u^2 + v^2}(u, v)$, where C_d is the drag coefficient and is determined by the following equation:

$$C_d = \max\left(\frac{k^2}{\ln\left(\frac{z_{ab}}{z_0}\right)^2}, 0.0025\right), \tag{8}$$

where k is the von Karman constant (0.4), z_0 is the bottom roughness parameter, and z_{ab} is the height above the bottom.

3.2. Model Setup

The model mesh covers the entire Gulf of Mexico, horizontally from 80.7° W to 97.9° W, and zonally from 18.1° N to 30.7° N (Figure 1). There are two open boundaries. One is located in the Caribbean Sea, connecting the east border of the Mexico and the west edge of Cuba. The other open boundary is located at the North Atlantic Ocean with the north point in the edge of Florida and the south point at the border of Cuba. The mesh contains 119,566 nodes and 214,297 elements. The finest resolution is about 50 m. There are 40 sigma layers. The model is three-dimensional barotropic, so salinity and temperature are not simulated nor discussed. The model is cold started with a time step of 1 s. The output time interval is set to be 30 min. The time periods for cold fronts and Hurricane Barry are from 20 December 2013 to 30 January 2014 and 20 June to 30 July 2019, respectively. Open boundary is only forced by tides, a combination of 10 tidal constituents (M2, S2, N2, K2, K1, O1, P1, Q1, MF, and MM). It is predicted by a tide model called TMD [54]. Wind stress and air pressure at mean sea level forcing at surface are obtained from the global Climate Forecast System Reanalysis (CFSR: https://climatedataguide.ucar.edu/climate-data/climate-forecast-system-reanalysis-cfsr) data with horizontal resolution of 0.5 degree by 0.5 degree.

3.3. Model Validation

The skill scores of FVCOM in simulating water elevation and along-channel velocity are 0.7 and 0.67 (Figure 2), which shows "excellent" performance based on Wu et al. [55]. The low-pass filtered water elevation and velocity are also in line with the observed data, which are categorized as "very good" with the skill scores of 0.51 and 0.60 (Figure 2). Discrepancies between the modeled and observed low-pass filtered water levels and along channel velocities may be caused by the uncertainty of bathymetry of these highly active tidal inlets. As mentioned earlier, tidal passes of Barataria bay are severely eroded. The model bathymetry may therefore have errors, leading to larger uncertainties in model results. Since our focus is on the weather-induced hydrodynamics, variables to be examined below are all low-pass filtered with a cut off frequency of 0.6 cycles per day.

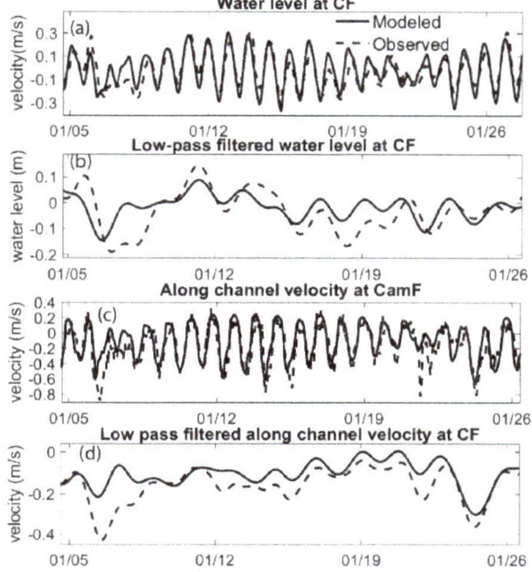

Figure 2. Validation of water level and along-channel velocity using observation in January 2014 at Caminada Pass, (**a,c**) are validations of water elevation and surface flow at Caminada Pass, (**b,d**) are validations of low-pass filtered water elevation and along channel velocity at Caminada Pass. Skill score for water level and along channel simulation are 0.7 and 0.67, which are categorized as "excellent".

3.4. Atmospheric Background

Here, we examine two types of atmospheric systems with different scales and influence regions. One is a cold front that entered the study area around 0000 UTC, 6 January 2014 and left the region around 1800 UTC, 6 January 2014. As shown in Figure 3a,b, the cold front developed from a low-pressure center (1003 hPa), which was located in Indiana. Southwesterly wind was dominating in our study site. As the front moved to the east, air pressure continued to drop to a minimum of 992 hPa. When the front was passing Barataria Bay, the southwesterly wind abruptly changed to northwesterly wind. After the cold front's passage, wind was from the northern quadrants for about two days with a maximum magnitude of 13.5 m/s (Figure 4a). This cold front passage is a typical weather phenomenon between late fall and the following spring (mostly October to April). During each of the frontal events, wind changes its direction from southern quadrants to northern quadrants when the cold front passages [50].

The other type of weather system studied here is Hurricane Barry. Hurricane Barry was first originated from a mesoscale convective vortex on 2 July 2019. It went into Gulf of Mexico on 10 July and developed into a tropical depression, before being upgraded to Tropical Storm Barry. It made its landfall on 13 July on Marsh Island, Louisiana, 190 km west of Barataria Bay as a Category 1 hurricane with a minimum sea-level pressure of 993 hPa and a maximum wind speed of 33 m/s (Figure 3c,d). The maximum wind speed measured at the Grand Isle station near Barataria Bay reached 14.5 m/s. Before Barry's landfall, northerly wind was dominating, and after the landfall of Barry, wind changed its direction to southerly wind and persisted for at least 4 days. From Figure 4b, one can see that Hurricane Barry had longer influencing period than the cold front mentioned above. Wind magnitude was also larger than that during the cold front.

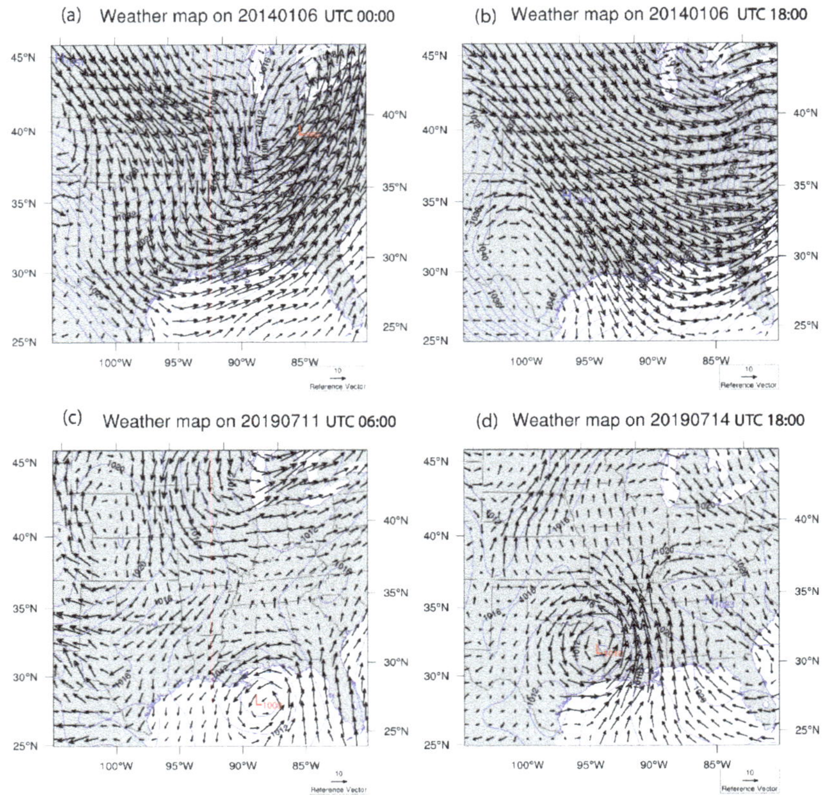

Figure 3. Weather maps during the cold front at 0000 UTC, 01/06 and 1800 UTC, 01/06, and during Hurricane Barry at 0600, 07/11 and 1800, 07/14. (**a**,**b**) are the weather maps before and after the cold front passage in January 2014, and (**c**,**d**) are weather maps before and after the landfall of Hurricane Barry. Blue lines are contours of air pressure.

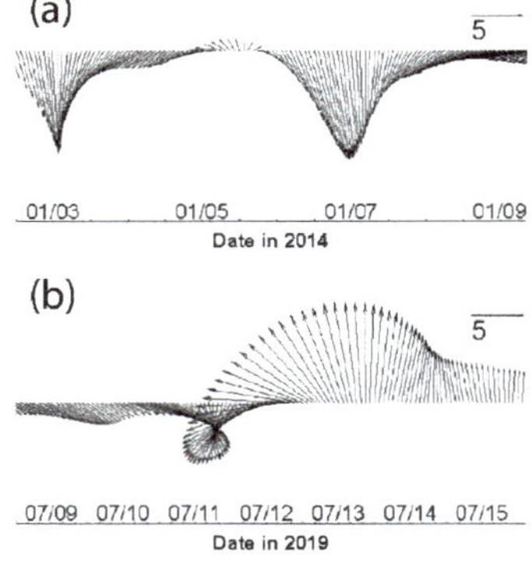

Figure 4. Wind vectors during cold front event in January 2014 and Hurricane Barry in July 2019. (max wind magnitude: 13.5 m/s and 14.5 m/s). (**a**,**b**) show the wind vectors during cold front and hurricane, respectively.

4. Results

4.1. Hydrodynamic Response

Low-pass filtered water level variations during cold front and hurricane are shown in Figure 5a,b, respectively. Water level dropped about 15 cm after the cold front passage during northerly wind, so that one can see an obvious trough at 1500 UTC on 6 January. This is because northerly wind after the cold front passage continuously blew the water out of the bay, leading to the water level minimum. As northerly wind weakened, water level begun to rise. On the other hand, when the water level was under the influence of Hurricane Barry, it is found that there was a surge of 40 cm after the hurricane's landfall, when wind also reached the maximum of 14.5 m/s. The surge resulted from the southerly wind after Barry's landfall, which blew the water from the coastal ocean into the bay. Compared with the trough induced by the cold front, the surge caused by Hurricane Barry is about 3 times larger in terms of magnitude.

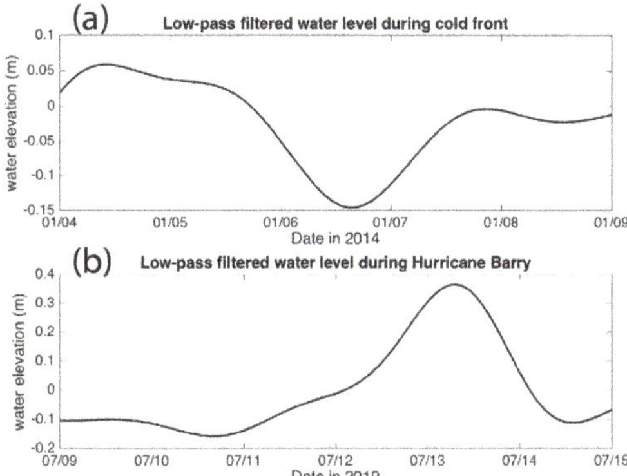

Figure 5. Water level variation during different weather types. (**a**) a cold front event and (**b**) Hurricane Barry.

Figure 6 shows the surface (Figure 6a) and bottom (Figure 6b) flows at 1200 UTC on 5 January 2014 before cold front passage when southerly wind dominated (with maximum magnitude of about 3 m/s). Flows inside the bay were in the wind's direction, flowing from south to north. Shallower and surface water had a larger magnitude of flows. Surface and bottom flows were unidirectional, while surface flow had a larger magnitude. After the cold front's passage (Figure 6c,d), when northerly wind dominated (maximum magnitude of 13.5 m/s), flows inside the bay were also in the direction of the wind, flowing from north to south. However, there existed a strong return flow in the middle of the bay where it had a greater water depth, which is consistent with previous studies in many systems: currents for shallower water tends to move in the direction of wind, whereas against the wind's direction for the region with a greater water depth (e.g., [49,56]). Again, surface and bottom flows were mostly uniform, except that surface flow had a larger magnitude. This is because bottom flow is decreased by bottom friction.

Figure 7 shows the surface (Figure 7a,c) and bottom (Figure 7b,d) flows prior (Figure 7a,b) and after (Figure 7c,d) Hurricane Barry's landfall. During Hurricane Barry, wind was rotating clockwise. Before Hurricane Barry's landfall, wind in the northern quadrant was dominating (with magnitude of 10 m/s). As a consequence, both surface and bottom currents were flowing in the direction of wind, from north to south in the shallow water region. An apparent returning flow against the wind's direction occurred in the deeper water region. After the landfall of Hurricane Barry, southerly wind dominated with the magnitude reaching 14.5 m/s. The surface and bottom flow in shallower water were flowing in the direction of the wind, moving from south to north, while in the deeper water region, they were flowing against the direction of wind, moving from north to south. Similarly, surface and bottom flows were mostly unidirectional, except that the surface flow had a larger magnitude.

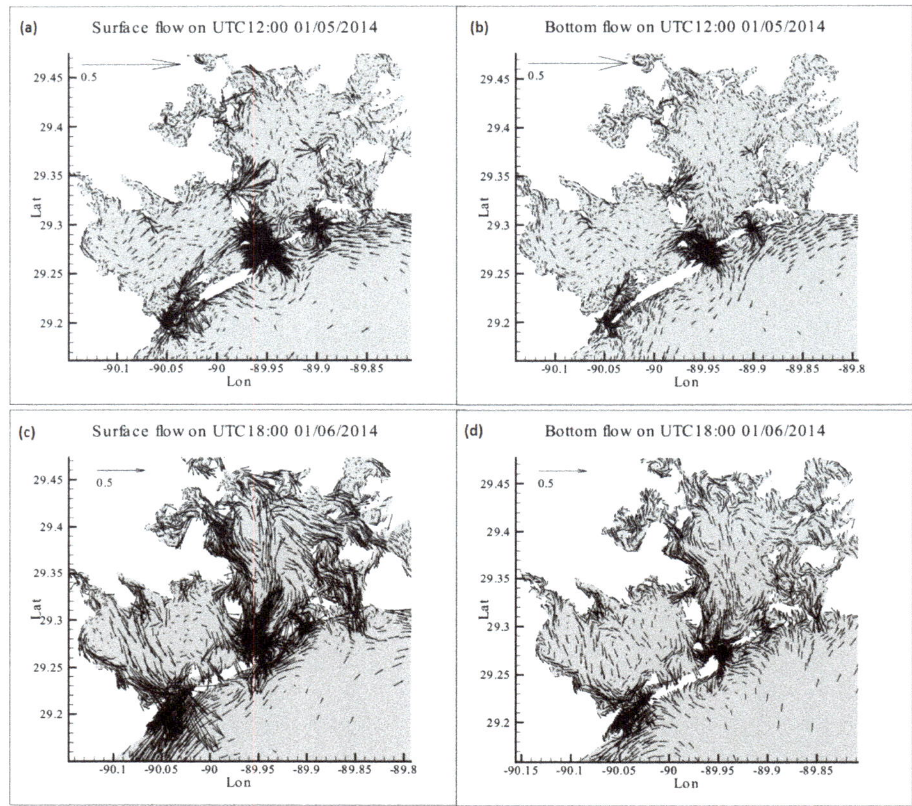

Figure 6. Surface and bottom flows influenced by the cold front event in January 2014. (**a**,**b**) are surface and bottom flows on 1200 UTC, 5 January 2014; (**c**,**d**) are surface and bottom flows on 1800 UTC, 1 January 2014.

4.2. Water Volume Transport

To examine the water transport through multiple inlets, water volume fluxes are calculated using the following equation [50]:

$$V(t)|_\Gamma = \int_{-H}^{\varsigma} (\int_0^L V_n(x,y,z,t)|_\Gamma d\xi)dz, \tag{9}$$

where V is the water volume flux in cubic meters per second. Γ is the transect perpendicular to the along-channel direction, H is the water depth, and ζ is the surface elevation. $V_n(x,y,z,t)$ is the low-pass filtered along-channel velocity in different water depths. A positive sign means water is transported into Barataria Bay.

Figure 8 shows the volume flux through Caminada Pass, Barataria Pass, Pass Abel, and Quatre Bayou. Positive value means water is flowing into the bay, while negative values means water is flowing out of the bay. During the cold front event in January 2014, the results indicate that water volume flux through Barataria Pass was the largest. Before cold front passage at 0000 on 6 January, water was flowing inside the bay through Caminada Pass and Quatre Bayou, while it was transported out of the bay through Barataria Pass and Pass Abel, which is consistent with mode 2a in Li et al. [50] under southerly wind. After the cold front's passage, as wind changed its direction from the south quadrants to north quadrants, volume flux through the four inlets began to decrease, then started to flow into the bay. Interestingly, volume flux changed the sign at different time stages: volume flux through Barataria Pass changed its sign first, followed by Pass Abel, Caminada Pass, and Quatre Bayou at last, which means there was a period (from 2000 UTC on 6 January to 0000 UTC on 7 January) when water was flowing out of the bay through Barataria Pass, while flowing inside of the bay through the other three inlets. From 0000 UTC on 7 January, water was transported into the bay through both Barataria Pass and Pass Abel, but it transported out of the bay through the other inlets, which is consistent with mode 2b in Li et al. [50] under northerly wind.

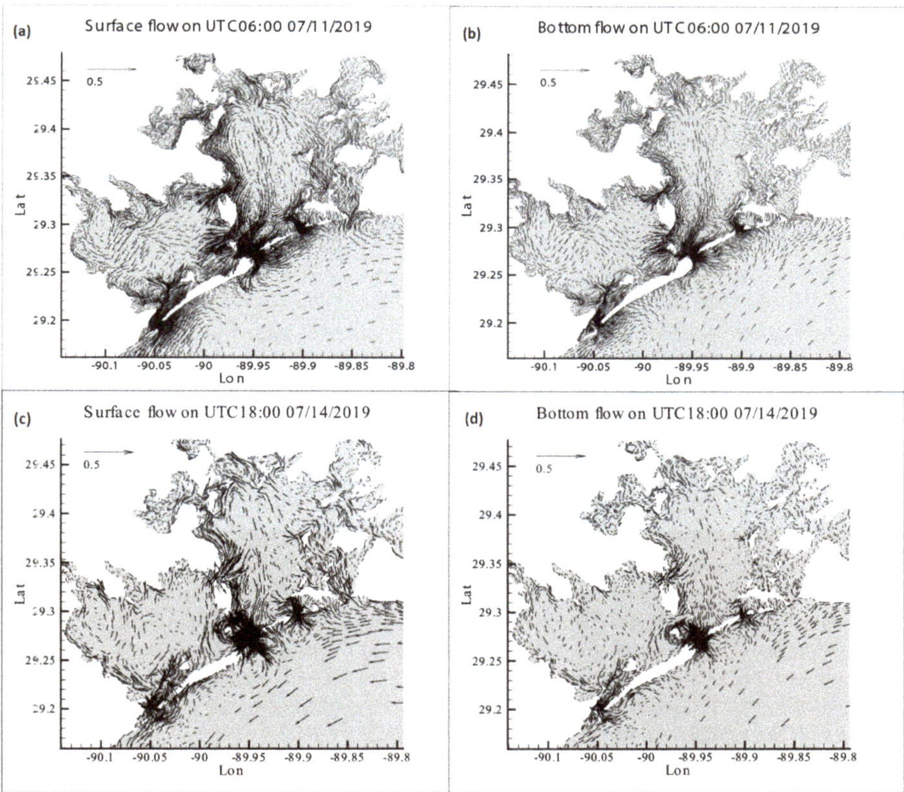

Figure 7. Surface and bottom flows under Hurricane Barry in 2019. (**a**,**b**) are surface and bottom flows on 0600 UTC, 11 July 2019; (**c**,**d**) are surface and bottom flows on 1800 UTC, 14 July 2019.

During Hurricane Barry, wind changed direction from the northern quadrants to southern quadrants. Therefore, the volume flux changed its sign after Barry's landfall on 13 July 2019, flowing into the bay through Barataria Pass, Pass Abel, and Quatre Bayou at first, then flowing out of the bay through these inlets afterwards. Note that Caminada Pass had the opposite condition: water was flowing out of the bay and then into the bay after Barry's landfall. This pattern is not included in any mode in cold front-induced flows [50]. Obviously, the mode of hurricane-induced flows can be different from that due to cold fronts. Before Barry's landfall, Pass Abel had the largest inward flux under northerly wind, and after Barry's landfall, Barataria Pass had the largest outward flux with a magnitude reaching 2500 m^3/s, which was about 4 times of that induced by the cold front.

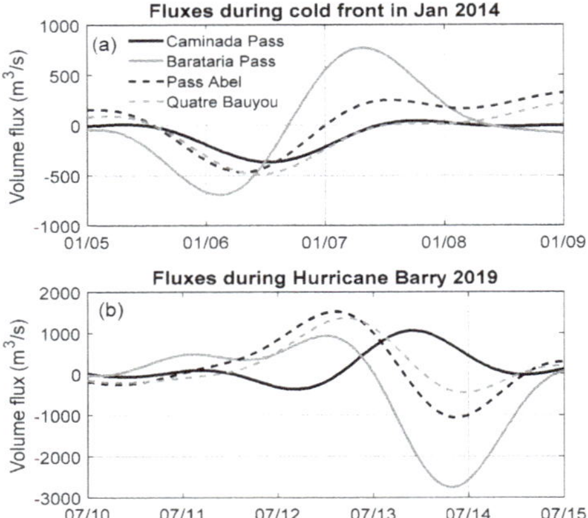

Figure 8. Volume flux through four inlets. (**a**) are fluxes in four passes during cold front, (**b**) are fluxes in four passes during Hurricane Barry.

5. Conclusions

Cold fronts and hurricanes have different wind patterns, and temporal and spatial characteristics: cold fronts usually have abrupt changes in wind direction (from southern quadrants to northern quadrants) before and after the frontal passage, with several hours to 3 days in duration, and 2000–3000 km length of front and a narrow width (less than 10 km) of the front. However, a hurricane has a radius of maximum wind (30–50 km) larger than the width of the frontal zone, and with a longer impact period of about a few days to a week, stronger maximum wind, and a rotating wind. Due to these differences, a cold front is more likely to result in a trough in water level post front, while hurricane tends to produce a more significant storm surge even severe inundation. The maximum of the variation in water level caused by Hurricane Barry was about 3 times of that caused by a cold front. The ratio is variable at different locations relative to the location of landfall.

Surface and bottom flows inside the Barataria bay have unidirectional movement except that surface flow has a larger magnitude. Currents tend to flow in the direction of wind in the shallower water region, but against the direction of wind in the deeper water region, resulting in a returning flow in the middle of bay. Water is transported out of the bay after a cold front passage. However, it is transported into the bay after Hurricane Barry's landfall. Volume flux through four inlets of Barataria Pass follows two EOF modes found by Li et al. [50] under a cold front event, in which water tends to be transported out of (into) the bay through Barataria Pass and Pass Abel during southerly (northerly) wind and into (out of) the bay through Caminada Pass and Quatre Bayou under southerly (northerly) wind. However, the volume flux through the four inlets of Barataria Bay shows different pattern under Hurricane Barry: water is flowing into (out of) the bay through Barataria Pass, Pass Abel, and Quatre Bayou under northerly (southerly) wind. The flux is the opposite through Caminada Pass, which is flowing out of (into) the bay during southerly (northerly) wind. The maximum water volume flux induced by hurricane is about 4 times that induced by a cold front event, indicating that the influence of four cold fronts is comparable with one hurricane event.

Author Contributions: Methodology, C.L.; validation, W.H.; formal analysis, W.H.; writing—original draft preparation, W.H.; writing—review and editing, C.L.; project administration, C.L. All authors have read and agreed to the published version of the manuscript.

Funding: The study was supported by National Science Foundation and Natural Environment Research Council (NSF-NERC 1736713), Bureau of Ocean Energy Management (M15AC000015), and GCOOS through NOAA-NOS-IOOS-2016-2004378.

Acknowledgments: The numerical modeling was performed on the cluster computers of the Louisiana Optical Network Initiative (LONI) HPC systems at LSU.

Conflicts of Interest: The authors declare no conflict of interest.

References

1. Penland, S.; Roberts, H.H.; Williams, S.J.; Sallenger, A.H., Jr.; Cahoon, D.R.; Davis, D.W.; Groat, C.G. Coastal Land Loss in Louisiana. In *Gulf Coast Association of Geological Societies Transactions*; GCAGS: Lafayette, LA, USA, 1990; Volume 40, pp. 685–699.

2. Britsch, L.D.; Dunbar, J.B. Land loss rates: Louisiana coastal plain. *J. Coast. Res.* **1993**, *9*, 324–338.

3. Roberts, H.H. Dynamic Changes of the Holocene Mississippi River Delta Plain: The Delta Cycle. *J. Coast. Res.* **1997**, *13*, 605–627.

4. Barras, J.; Beville, S.; Britsch, D.; Hartley, S.; Hawes, S.; Johnston, J.; Kemp, P.; Kinler, Q.; Martucci, A.; Porthouse, J.; et al. Historical and Projected Coastal Louisiana Land Changes: 1978–2050. In *U.S. Geological Survey Scientific Investigations*; U.S. Geological Survey: Reston, VA, USA, 2003.

5. Penland, S.; Connor, P.F., Jr.; Beall, A.; Fearnley, S.; Williams, S.J. Changes in Louisiana's Shoreline: 1855–2002. *J. Coast. Res.* **2005**, *44*, 7–39.

6. Palaseanu-Lovejoy, M.; Kranenburg, C.; Barras, J.A.; Brock, J.C. Land Loss Due to Recent Hurricanes in Coastal Louisiana, USA. *J. Coast. Res.* **2013**, *63*, 97–109. [CrossRef]

7. Harris, D.L. Tides and Tidal datums in the United States. In *Special Reports*; U.S. Army Corps of Engineers: Washington, DC, USA, 1981; Volume 7, p. 382.

8. Forbes, M.J. Hydrologic investigations of the lower Calcasieu River, Louisiana. In *Water Resources Investigations Report 87-4173*; U.S. Geological Survey: Reston, VA, USA, 1988; p. 61.

9. Stone, G.W.; Liu, B.; Pepper, D.A.; Wang, P. The importance of extratropical and tropical cyclones on the short-term evolution of barrier islands along the northern Gulf of Mexico, USA. *Mar. Geol.* **2004**, *210*, 63–78. [CrossRef]

10. Khalil, S.M.; Lee, D.M. Restoration of isles dernieres, Louisiana: Some reflections on morphodynamic approaches in the Northern Gulf of Mexico to conserve coastal/marine systems. *J. Coast. Res.* **2006**, *39*, 65–71.

11. Day, J.W., Jr.; Boesch, D.F.; Clairain, E.J.; Kemp, G.P.; Laska, S.B.; Mitsch, W.J.; Orth, K.; Mashriqui, H.; Reed, D.J.; Shabman, L.; et al. Restoration of the Mississippi Delta: Lessons from Hurricanes Katrina and Rita. *Science* **2007**, *315*, 1679–1684. [CrossRef]

12. Morton, R.A. Historical Changes in the Mississippi-Alabama Barrier-Island Chain and the Roles of Extreme Storms, Sea Level, and Human Activities. *J. Coast. Res.* **2008**, *24*, 1587–1600. [CrossRef]

13. Otvos, E.G.; Carter, G.A. Hurricane Degradation-Barrier Development Cycles, Northeastern Gulf of Mexico: Landform Evolution and Island Chain History. *J. Coast. Res.* **2008**, *24*, 463–478. [CrossRef]

14. Morton, R.A.; Barras, J.A. Hurricane Impacts on Coastal Wetlands: A Half-Century Record of Storm-Generated Features from Southern Louisiana. *J. Coast. Res.* **2011**, *27*, 27–43. [CrossRef]

15. Dingler, J.R.; Reiss, T.E.; Plant, N.G. Erosional Patterns of the Isles Dernieres, Louisiana, in Relation to Meteorological Influences. *J. Coast. Res.* **1993**, *9*, 112–125.

16. Houston, S.H.; Shaffer, W.A.; Powell, M.D.; Chen, J. Comparisons of HRD and SLOSH surface wind fields in hurricanes: Implications for storm surge modeling. *Weather Forecast.* **1999**, *14*, 671–686. [CrossRef]

17. Hubbert, G.D.; McInnes, K.L. A storm surge model for coastal planning and impact studies. *J. Coast. Res.* **1999**, *15*, 168–185.

18. Peng, M.; Xie, L.; Pietrafesa, L.J. A numerical study of storm surge and inundation in the Croatan–Albemarle–Pamlico Estuary System. *Estuar. Coast. Shelf Sci.* **2004**, *59*, 121–137. [CrossRef]

19. Peng, M.; Xie, L.; Pietrafesa, L.J. Tropical cyclone induced asymmetry of sea level surge and fall and its presentation in a storm surge model with parametric wind fields. *Ocean Model.* **2006**, *14*, 81–101. [CrossRef]

20. Xia, M.; Xie, L.; Pietrafesa, L.J.; Peng, M. A numerical study of storm surge in the Cape Fear River Estuary and adjacent coast. *J. Coast. Res.* **2008**, *24*, 159–167. [CrossRef]

21. Chuang, W.S.; Wiseman, W.J., Jr. Coastal sea level response to frontal passages on the Louisiana–Texas Shelf. *J. Coast. Res.* **1983**, *88*, 2615–2620. [CrossRef]

22. Siadatmousavi, S.M.; Jose, F. Winter storm-induced hydrodynamics and morphological response of a shallow transgressive shoal complex: Northern Gulf of Mexico. *Estuar. Coast. Shelf Sci.* **2015**, *154*, 58–68. [CrossRef]

23. Walker, N.D.; Hammack, A.B. Impacts of Winter Storms on Circulation and Sediment Transport: Atchafalaya-Vermilion Bay Region, Louisiana, USA. *J. Coast. Res.* **2000**, *16*, 996–1010.

24. Steyer, G.D.; Perez, B.C.; Piazza, S.; Suir, G. Potential consequences of saltwater intrusion associated with hurricanes Katrina and Rita. *Sci. Storms* **2005**, *2007*, 137–145.

25. Mulrennan, M.E.; Woodroffe, C. Saltwater intrusion into the coastal plains of the Lower Mary River, Northern Territory, Australia. *J. Environ. Manag.* **1998**, *54*, 169–188. [CrossRef]

26. Pezeshki, S.R.; Delaune, R.D.; Patrick, W.H., Jr. Flooding and saltwater intrusion: Potential effects on survival and productivity of wetland forests along the U.S. Gulf Coast. *For. Ecol. Manag.* **1990**, *33–34*, 287–301. [CrossRef]

27. Neubauer, S.C. Ecosystem Responses of a Tidal Freshwater Marsh Experiencing Saltwater Intrusion and Altered Hydrology. *Estuaries Coasts* **2013**, *36*, 491–507. [CrossRef]

28. Xue, L.; Li, X.; Yan, Z.; Zhang, Q.; Ding, W.; Huang, X.; Tian, B.; Ge, Z.; Yin, Q. Native and non-native halophytes resiliency against sea-level rise and saltwater intrusion. *Hydrobiologia* **2018**, *806*, 47–65. [CrossRef]

29. Fritz, H.M.; Blount, C.; Sokoloski, R.; Singleton, J.; Fuggle, A.; McAdoo, B.G.; Moore, A.; Grass, C.; Tate, B. Hurricane Katrina storm surge distribution and field observations on the Mississippi Barrier Islands. *Estuar. Coast. Shelf Sci.* **2007**, *74*, 12–20. [CrossRef]

30. Williams, H.F.L. Stratigraphy, Sedimentology, and Microfossil Content of Hurricane Rita Storm Surge Deposits in Southwest Louisiana. *J. Coast. Res.* **2009**, *25*, 1041–1051. [CrossRef]

31. Williams, H.F.L.; Flanagan, W.M. Contribution of Hurricane Rita Storm Surge Deposition to Long-Term Sedimentation in Louisiana Coastal Woodlands and Marshes. *J. Coast. Res.* **2009**, *56*, 1671–1675.

32. Ebersole, B.A.; Westerink, J.J.; Bunya, S.; Dietrich, J.C.; Cialone, M.A. Development of storm surge which led to flooding in St. Bernard Polder during Hurricane Katrina. *Ocean Eng.* **2010**, *37*, 91–103. [CrossRef]

33. Niedoroda, A.W.; Resio, D.T.; Toro, G.R.; Divoky, D.; Das, H.S.; Reed, C.W. Analysis of the coastal Mississippi storm surge hazard. *Ocean Eng.* **2010**, *37*, 82–90. [CrossRef]

34. Rego, J.L.; Li, C. On the importance of the forward speed of hurricanes in storm surge forecasting: A numerical study. *Geophys. Res. Lett.* **2009**, *36*, L07609. [CrossRef]

35. Rego, J.L.; Li, C. On the receding of storm surge along Louisiana's low-lying coast. *J. Coast. Res.* **2009**, *SI56*, 1045–1049.

36. Li, C.; Weeks, E.; Rego, J.L. In situ measurements of saltwater flux through tidal passes of Lake Pontchartrain estuary by Hurricanes Gustav and Ike in September 2008. *Geophys. Res. Lett.* **2009**, *36*, L19609. [CrossRef]

37. Li, C.; Weeks, E.; Blanchard, B.W. Storm surge induced flux through multiple tidal passes of Lake Pontchartrain estuary during Hurricanes Gustav and Ike. *Estuar. Coast. Shelf Sci.* **2010**, *87*, 517–525. [CrossRef]

38. Shuckburgh, E.; Mitchell, D.; Stott, P. Hurricanes Harvey, Irma and Maria: How natural were these 'natural disasters'? *Weather* **2017**, *72*, 353–354. [CrossRef]

39. Li, C.; Weeks, E.; Milan, B.; Huang, W.; Wu, R. Weather Induced Transport through a Tidal Channel Calibrated by an Unmanned Boat. *J. Atmos. Ocean. Technol.* **2018**, *35*, 261–279. [CrossRef]

40. Keen, T.R. Waves and currents during a winter cold front in the Mississippi Bight, Gulf of Mexico: Implications for barrier island erosion. *J. Coast. Res.* **2002**, *18*, 6220636.

41. Keen, T.R.; Stavn, R.H. Hydrodynamics and marine optics during cold fronts at Santa Rosa Island, Florida. *J. Coast. Res.* **2012**, *28*, 1073–1087. [CrossRef]

42. Warner, J.C.; Butman, B.; Dalyander, P.S. Storm-driven sediment transport in Massachusetts Bay. *Cont. Shelf Res.* **2008**, *28*, 257–282. [CrossRef]

43. Perez, B.C.; Day, J.W., Jr.; Rouse, L.J.; Shaw, R.F.; Wang, M. Influence of Atchafalaya River Discharge and Winter Frontal Passage on Suspended Sediment Concentration and Flux in Fourleague Bay, Louisiana. *Estuarine Coast. Shelf Sci.* **2000**, *50*, 271–290. [CrossRef]

44. Kineke, G.C.; Higgins, E.E.; Hart, K.; Velasco, D. Fine-sediment Transport Associated with Cold-front passages on the shallow shelf, Gulf of Mexico. *Cont. Shelf Res.* **2006**, *26*, 2073–2091. [CrossRef]

45. Siadatmousavi, S.M.; Allahdadi, M.N.; Chen, Q.; Jose, F.; Roberts, H.H. Simulation of wave damping during a cold front over the muddy Atchafalaya Shelf. *Cont. Shelf Res.* **2012**, *47*, 165–177. [CrossRef]

46. Henrie, K.; Valle-Levinson, A. Subtidal variability in water levels inside a subtropical estuary. *J. Geophys. Res. Oceans* **2014**, *119*, 7483–7492. [CrossRef]

47. Murphy, P.; Waterhouse, A.F.; Hesser, T.J.; Penko, A.M.; Valle-Levinson, A. Subtidal flow and its variability at the entrance to a subtropical lagoon. *Cont. Shelf Res.* **2009**, *29*, 2318–2332. [CrossRef]

48. Sepulveda, H.H.; Valle-Levinson, A.; Framinan, M.B. Observations of subtidal and tidal flow in the Rio de la Plata Estuary. *Cont. Shelf Res.* **2004**, *24*, 509–525. [CrossRef]

49. Lin, J.; Li, C.; Boswell, K.; Kimball, M.; Rozas, L. Examination of winter circulation in a northern Gulf of Mexico estuary. *Estuaries Coasts* **2016**, *39*, 879–899. [CrossRef]

50. Li, C.; Huang, W.; Milan, B. Atmospheric Cold Front Induced Exchange Flows through a Microtidal Multi-inlet Bay: Analysis using Multiple Horizontal ADCPs and FVCOM Simulations. *J. Atmos. Ocean. Technol.* **2019**, *36*, 443–472. [CrossRef]

51. Li, C.; Huang, W.; Wu, R.; Sheremet, A. Weather induced quasi-periodic motions in estuaries and bays: Meteorological tide. *China Ocean Eng.* **2020**, *34*, 299–313. [CrossRef]

52. Chen, C.; Liu, H.; Beardsley, R.C. An unstructured grid, finite-volume, three dimensional, primitive equation ocean model: Application to coastal ocean and estuaries. *J. Atmos. Ocean. Technol.* **2003**, *20*, 159–186. [CrossRef]

53. Mellor, G.L.; Yamada, T. Development of a turbulence closure model for geophysical fluid problem. *Rev. Geophys. Space Phys.* **1982**, *20*, 851–875. [CrossRef]

54. Padman, L.; Erofeeva, S. A Barotropic Inverse Tidal Model for the Arctic Ocean. *Geophys. Res. Lett.* **2004**, *31*. [CrossRef]

55. Wu, H.; Zhu, J.R.; Shen, J.; Wang, H. Tidal modulation on the Changjiang River plume in summer. *J. Geophys. Res.* **2011**, *116*. [CrossRef]

56. Huang, W.; Li, C. Cold Front Driven Flows Through Multiple Inlets of Lake Pontchartrain Estuary. *J. Geophys. Res. Ocean.* **2017**, *122*, 8627–8645. [CrossRef]

Publisher's Note: MDPI stays neutral with regard to jurisdictional claims in published maps and institutional affiliations.

Journal of
*Marine Science
and Engineering*

Article

Data-Driven, Multi-Model Workflow Suggests Strong Influence from Hurricanes on the Generation of Turbidity Currents in the Gulf of Mexico

Courtney K. Harris [1,*], Jaia Syvitski [2], H.G. Arango [3], E.H. Meiburg [4], Sagy Cohen [5], C.J. Jenkins [2], Justin J. Birchler [6], E.W.H. Hutton [2], T.A. Kniskern [1], S. Radhakrishnan [4] and Guillermo Auad [7]

[1] Department of Physical Sciences, Virginia Institute of Marine Science, William & Mary, Williamsburg, VA 23062, USA; tara.kniskern@gmail.com
[2] CSDMS/INSTAAR, University of Colorado, Boulder, CO 80305, USA; jai.syvitski@Colorado.EDU (J.S.); jenkinsc0@gmail.com (C.J.J.); Eric.Hutton@colorado.edu (E.W.H.H.)
[3] Department of Marine and Coastal Sciences, Rutgers University, New Brunswick, NJ 08901, USA; arango@marine.rutgers.edu
[4] Department of Mechanical Engineering, University of California, Santa Barbara, CA 93106, USA; meiburg@engineering.ucsb.edu (E.H.M.); senthilradh@gmail.com (S.R.)
[5] Department of Geography, University of Alabama, Tuscaloosa, AL 35487, USA; sagy.cohen@ua.edu
[6] U.S. Geological Survey, Saint Petersburg, FL 33701, USA; jbirchler@usgs.gov
[7] Bureau of Ocean Energy Management, Sterling, VA 20166, USA; guillermo.auad@boem.gov
* Correspondence: ckharris@vims.edu

Received: 20 June 2020; Accepted: 25 July 2020; Published: 6 August 2020

Abstract: Turbidity currents deliver sediment rapidly from the continental shelf to the slope and beyond; and can be triggered by processes such as shelf resuspension during oceanic storms; mass failure of slope deposits due to sediment- and wave-pressure loadings; and localized events that grow into sustained currents via self-amplifying ignition. Because these operate over multiple spatial and temporal scales, ranging from the eddy-scale to continental-scale; coupled numerical models that represent the full transport pathway have proved elusive though individual models have been developed to describe each of these processes. Toward a more holistic tool, a numerical workflow was developed to address pathways for sediment routing from terrestrial and coastal sources, across the continental shelf and ultimately down continental slope canyons of the northern Gulf of Mexico, where offshore infrastructure is susceptible to damage by turbidity currents. Workflow components included: (1) a calibrated simulator for fluvial discharge (Water Balance Model - Sediment; *WBMsed*); (2) domain grids for seabed sediment textures (*dbSEABED*); bathymetry, and channelization; (3) a simulator for ocean dynamics and resuspension (the Regional Ocean Modeling System; *ROMS*); (4) A simulator (*HurriSlip*) of seafloor failure and flow ignition; and (5) A Reynolds-averaged Navier–Stokes (*RANS*) turbidity current model (*TURBINS*). Model simulations explored physical oceanic conditions that might generate turbidity currents, and allowed the workflow to be tested for a year that included two hurricanes. Results showed that extreme storms were especially effective at delivering sediment from coastal source areas to the deep sea, at timescales that ranged from individual wave events (~hours), to the settling lag of fine sediment (~days).

Keywords: turbidity current; suspended sediment; numerical model; Gulf of Mexico

---

1. Introduction

The Gulf of Mexico continental margin generates >1.7 million barrels of oil per day, through >3500 oil platforms. The northern Gulf of Mexico houses >45,000 km of underwater pipes that may be exposed to structural damage from extreme oceanic events. During the passage of a hurricane, storm waves can exceed 10 m in height, resuspending seafloor sediment and potentially liquefying the seafloor. Both of these mechanisms may induce sediment turbidity currents, and in fact, ~5% of the underwater petroleum pipes appear to be broken or damaged by sudden powerful turbidity currents (BOEM pers. comm. 2015). For example, in 2004, a large sediment failure in the wake of Hurricane Ivan toppled an oil platform offshore of the Gulf of Mexico and moved it ~0.17 km downslope, initiating oil and gas leaks at a water depth of 140 m [1]. Leakage from such offshore oil and gas infrastructure puts at risk about 40% of the USA's coastal and estuarine wetlands, which are vital to recreation, agriculture, and a $1B/y seafood industry [2].

Turbidity currents are important transport mechanisms in submarine canyons [3,4], such as the Mississippi and the De Soto Canyons, which incise the continental slope offshore of the Mississippi Delta. Several processes have been shown to have the potential to generate turbidity currents, including physical oceanographic mechanisms. Internal wave breaking on the upper slope may mobilize seafloor sediment [5]. Wave-current interactions on continental shelves during large oceanic storms can initiate wave-supported gravity flows [6]. Continental slope deposits may experience sediment failure triggered by sediment loading and over-steepening, and aided by excess pore pressure brought on by ground accelerations [7,8]. Localized events may grow into sustained currents via a self-amplifying 'ignition' process with accelerating erosion and entrainment of sediment from the seafloor [9,10].

While the relative importance of these mechanisms in the northern Gulf of Mexico remains to be seen, evidence points to the potential for oceanic storms to mobilize sediment there, either during the passage of moderate storms [11] or more extreme events such as hurricanes [12]. Analysis of sediment deposits indicated that most (~75%) of the sediment budget of the Mississippi Canyon could be attributed to delivery during major hurricanes, likely through gravity-driven transport [13]. Several processes affect the seafloor during short-lived hurricane passages, including sediment mass failures, erosion, and suspension. For example, mudflows in the Mississippi Delta area, triggered by the 1969 Category 5 Hurricane Camille, destroyed the offshore platform SB-70B. The seafloor at a depth of about 90 m moved more than 1000 m downslope with soil flows up to 30 m in thickness [14]. Seafloor shear stresses from waves and currents of up to 1 N/m^2 were monitored at a depth of 90 m during the 2004 Category 5 Hurricane Ivan, reaching the critical shear stress for fine gravel [15]. The Ivan event lifted suspended sediment as high as 25 m in the water column and eroded the seafloor up to 0.30 m vertically over more than 500 km^2, thus removing hundreds of millions of tons of sediment with deposits at the shelf edge and upper slope [16], and additionally causing apparent damage to oil infrastructure [1]. Evidence of the effects of large storms at great depth in the Gulf of Mexico has been seen in conjunction with other hurricanes, such as Hurricane Georges in the Mississippi Canyon [12]; Hurricane Frederic in the De Soto Canyon [17]; and Hurricane Allen [18]. Rapid loading of sections of the seafloor locally enhances the prospects for gravitational slope failures, given the associated rapid increase in pore pressures and reduction in effective sediment strengths [7]. Process-based numerical modeling offers a way to study such ephemeral high-energy processes.

Studies of sediment dispersal on continental margins, including the northern Gulf of Mexico, have typically focused on an individual component of the transport path such as gravity-driven transfer via canyons, shelf resuspension, or flood plume dispersal. For example, numerical models for suspended sediment transport have been developed and applied to the northern Gulf of Mexico [19–22], but these types of suspended transport models have not been directly linked to turbidity current models. This paper describes a numerical capability to simulate the transport of sediment, from fluvial sources, to the continental shelf, the deeper continental slope, and ultimate depocenters. Accounting for these sediment transport pathways, and the hazards that they present, is a problem of multi-scale physics, ranging from continental-scale drainage basins that deliver sediment to the sea, to shelf-wide storm systems that mobilize and redistribute sediment, to small-scale turbulent motions that affect turbidity current generation and structure.

This paper describes a loosely coupled numerical workflow that has been developed to address land-sea pathways for sediment routing of terrestrial and coastal sources, across the continental shelf, and ultimately down the continental slope and canyons of the northern Gulf of Mexico. Few studies have attempted to integrate the various transport mechanisms into a single comprehensive framework, accounting for the multi-scale physics that are relevant to the full sediment transport pathway. The workflow was used to explore conditions that may trigger episodes of sediment transport onto the continental slope and to evaluate two hypotheses: (1) episodic sediment transport down a submarine canyon is fed by sediment input at the canyon head from wave and current resuspension, and (2) turbidity currents are triggered by failures near the shelf-slope break and are likely to pass into the canyons of the continental slope. Simulation results were based on oceanographic and meteorological conditions that could impact the generation of turbidity currents. The workflow (Figure 1) includes modules that:

(1) Simulate the fluvial delivery of water and sediment into the Gulf of Mexico with the Water Balance Model-Sediment (*WBMsed*) and as augmented by USGS (US Geological Survey) and USACE (US Army Corps of Engineers) gauged river data;

(2) Develop domain grids and bathymetry for ocean circulation and sediment transport models;

(3) Compute spatial griddings of seabed sediment texture from *dbSEABED*, and of topographic channelization from the bathymetry, for use in sediment transport and seabed failure models;

(4) Employ a high resolution (10 km) spectral wave action model (*WaveWatch III®)* driven by *GFDL-GFS* (Geophysical Fluid Dynamics Laboratory–Global Forecast System) winds for use in the ocean and sediment transport models;

(5) Calculate hourly-timescale ocean circulation at a spatial resolution of a few kilometers via the Regional Ocean Modeling System (*ROMS*) forced with *ECMWF* (European Centre for Medium-Range Weather Forecasts) *ERA* (ECMWF Re-Analysis) winds;

(6) Represent seafloor resuspension and transport at the same resolution as *ROMS'* hydrodynamics using the Community Sediment Transport Modeling System (*CSTMS*);

(7) Apply seabed mass-failure and a sediment suspension model (*HurriSlip*) to determine failure and ignition locations, and the conditions to be used as input to the turbidity current model;

(8) Develop and deploy a Reynolds-averaged Navier–Stokes (*RANS*) model (*TURBINS*) to route sediment flows down the Gulf of Mexico slopes and canyons, providing estimates of bottom shear stress needed for ascertaining possible damage to offshore infrastructure.

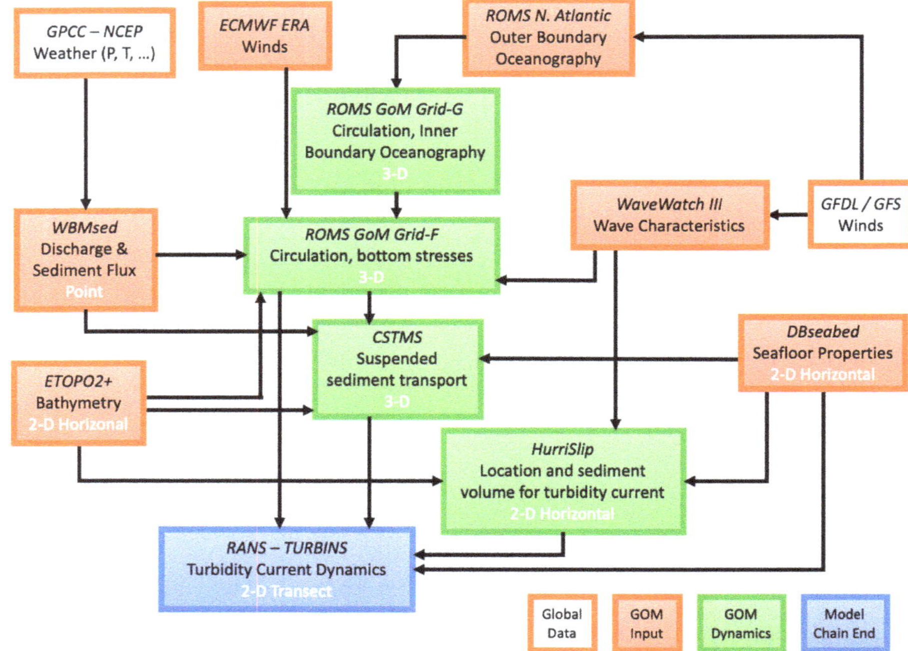

Figure 1. Workflow showing models employed and boundary data usage. Models and data systems discussed in detail in the text. The white text identifies whether the models ran were Point models, 2D horizontal plan-view; 2D vertical transect or 3D.

2. Materials and Methods

Section 2.1 describes the northern Gulf of Mexico, where the model workflow was applied. Section 2.2 provides descriptions and methods for each component of the workflow, noting how the various components can interact with one another. Section 2.3 outlines the implementation of the suite of models used to evaluate sediment routing in the northern Gulf of Mexico.

2.1. Environmental Setting

The Mississippi River drains 41% of the continental United States before entering the northern Gulf of Mexico (Figure 2). The discharge of the Mississippi River is regulated so that approximately 70% of it enters the Gulf through its main Mississippi River channel, while the remaining 30% enters through the Atchafalaya River channel [23]. Average modern-day sediment loads of the Mississippi and Atchafalaya Rivers are 115 and 57 Mt/yr, respectively [23]. Sand is deposited near the river mouths while most of the remaining suspended silts and muds are dispersed more widely [19,24,25]. Rapid delta progradation during the Holocene has narrowed and steepened the continental shelf (~20 km wide, ~0.4° gradient). The Mississippi Canyon, which cuts into the continental slope to the west of the bird-foot delta has been implicated as a conduit for shelf sediment during large storms [12,26].

A fair amount is known about suspended sediment dispersal on the Gulf of Mexico continental shelf. Frontal systems that occur frequently during winter months can create energetic waves and currents that cause significant sediment transport [27,28]. Wave contributions dominate the bed stresses on the continental shelf offshore of the Mississippi Delta, but fairweather waves are typically capable of mobilizing the seabed only in the surf and nearshore zones [19]. During extreme oceanic storms, however, deep-water wave heights exceed 10 m, with nearshore waves east of the bird-foot delta reaching 9 m in 15 m of water during Hurricane Ivan [29]. Storm waves, either from moderate storms or intense but infrequent hurricanes, have been shown to mobilize sediment mass failures on the Mississippi River Delta Front at water depths of ~75 m [11]. Sediment trap data and allied mooring and camera data from deep-water locations (~1000 m) have indicated that frequent, small magnitude resuspension events driven by inertial currents contribute to sediment transport there [30]. Less is known, however, about the mechanisms that drive shelf–slope sediment exchange or transport down the continental slope or canyons.

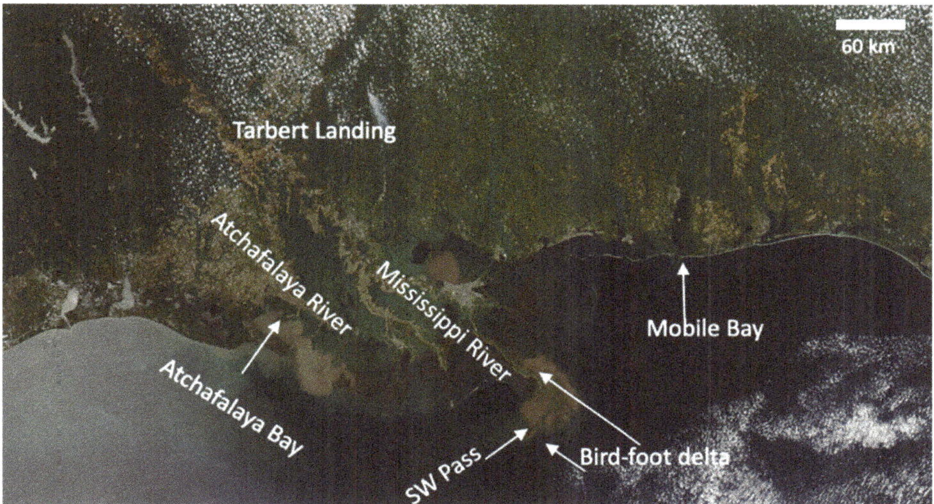

Figure 2. Study area identifying locations of bird-foot delta and Southwest (SW) Pass of Mississippi River; Atchafalaya River and Bay; Mobile Bay; Tarbert Landing (site of commonly used river gauge). Satellite image of Mary 17, 2011 from MODIS on NASA's Aqua satellite.

2.2. Workflow Components

Sections 2.2.1–2.2.6 describe individual workflow components, each developed to quantify a different component of the sediment dispersal pathway, from delivery of sediment to the norhtern Gulf of Mexico from river discharge, to turbidity current transport in deep water.

2.2.1. River Discharge Modeling Results and Observations

Few rivers that discharge into the Gulf are adequately gauged, with only the Mississippi and Pearl Rivers having associated sediment flux determinations. Therefore, a global *WBMsed* [31,32] was used to estimate daily discharge and sediment flux from rivers into the northern Gulf of Mexico. *WBMsed* combined the Water Balance Model (*WBM*) with the *BQART* and *Psi* models. Specifically, *BQART* simulates long-term (30+ years) average suspended sediment loads for a basin outlet and is based on individual upstream basin properties for each distributed pixel, including geographical, geological and human factors [33]. The *Psi* variability model resolves the suspended sediment flux on a daily time step from the long-term sediment flux estimated by *BQART*, able to capture the intra-annual and inter-annual variability observed in natural river systems [34]. Skill assessment of *WBMsed* was based on a comparison to daily USGS observations of water and sediment discharge [35]; and daily discharge predictions compared favorably to both ground-based gauging stations and satellite-based observations [31,36]. Sixteen rivers that discharge to the northern Gulf of Mexico (Figure 3A) were simulated using observed conditions for 1995–2011.

Freshwater discharge to the northern Gulf of Mexico peaks seasonally in February through March (Figure 3B). The combined Mississippi–Atchafalaya Rivers supply 81% of the freshwater discharge into the northern Gulf; other significant riverine sources are identified in Figure 3A. The combined Mississippi and Atchafalaya flow averages 20,874 m^3/s with a standard deviation of 11,211 m^3/s (USACE observations). The *WBMsed* estimate of the combined Atchafalaya–Mississippi discharge for the same period was 18,300 m^3/s, with a standard deviation of 12,400 m^3/s. The total predicted (1995–2011) discharge for all northern Gulf rivers was 22,800 m^3/s, with a standard deviation of 15,400 m^3/s. The merged discharge of non-Mississippi rivers was 4540 m^3/s or 19% of the total flow into the northern Gulf of Mexico, with a standard deviation of 4730 m^3/s (Figure 3C). The 17 y one-day high of these non-Mississippi sources was 30,000 m^3/s (8 March 1998), highlighting a potential issue of studies that solely consider the Mississippi River discharge. On 13 November 1997, these non-Mississippi sources accounted for 66% of the total freshwater input into the Gulf (of 15,500 m^3/s), a 17 y one-day maximum contribution.

The USGS observations (1995–2011) of Mississippi River sediment load indicate a mean value of 170 Mt/y. Sediment discharge to the northern Gulf of Mexico is seasonal but with peak loads of short duration (Figure 3D). On average, the Mississippi River supplies 88% of the fluvial sediment load to the northern Gulf, although its contribution varies from more than 99% to less than 15% on any given day. Based on *WBMsed*, the 14 non-Mississippi Rivers identified in Figure 3A supplied 16.2 Mt/y of sediment to the northern Gulf of Mexico during the same period.

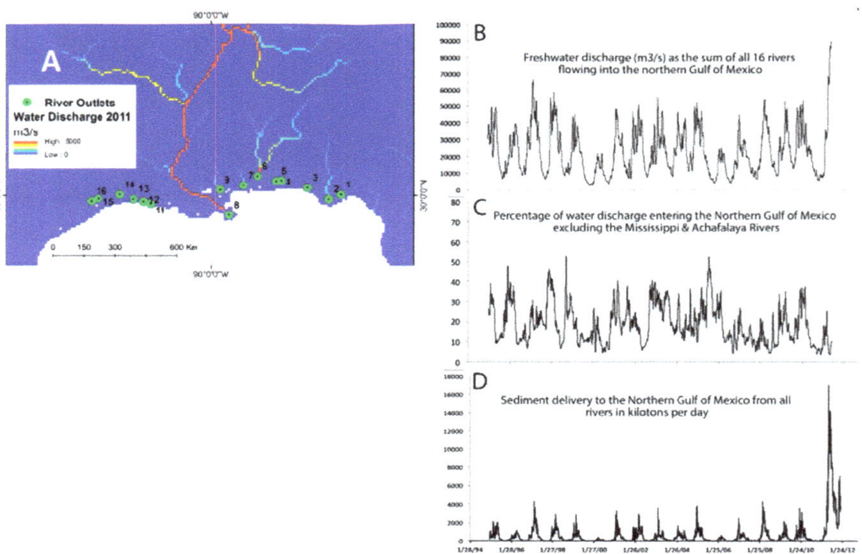

Figure 3. (**A**) *WBMsed* model: color shows discharge rates for March 1, 2005; numbers identify river outlets: (1) Apalachee, (2) Apalachicola, (3) Conecuh, (4) Choctawhatchee, (5) Escambia, (6) Mobile, (7) Pascagoula, (8) Mississippi and Atchafalaya, (9) Pearl, (11) Grand, (12) Sulphur, (13) Sabine, (14) Neches, (15) Trinity, and (16) Conroe. (**B**) Total daily water discharge (m^3/s) entering the northern Gulf of Mexico (*WBMsed* simulation). The 2011 flood season (partly shown) was amongst the most devastating floods in the continental US history. (**C**) The percentage of non-Mississippi freshwater discharge entering the northern Gulf of Mexico study region (*WBMsed* simulations). (**D**) Total suspended load (Kt/d) entering the northern Gulf of Mexico (*WBMsed* simulations).

2.2.2. Ocean Hydrodynamic and Wave Model

ROMS is a mature numerical framework that represents ocean dynamics over a wide range of spatial (coastal to basin) and temporal (days to inter-annual) scales. A three-dimensional, free-surface, terrain-following ocean model, *ROMS* resolves the primitive momentum and continuity equations modeling large-scale ocean circulation using the hydrostatic vertical momentum balance and Boussinesq approximation [37,38]. The dynamical kernel includes accurate and efficient algorithms for time-stepping, advection, pressure gradient [38,39], subgrid-scale parameterizations to represent small-scale turbulent processes [40,41] and various bottom boundary layer formulations to determine the stress exerted on the flow by the sediment bed.

For our implementation, two nested *ROMS* grids were run. The larger-scale coarser model represented hydrodynamics over the entire Gulf of Mexico (Figure 4, black box; hereafter, Grid-g) and provided boundary conditions to a finer-scale model that calculated higher-resolution hydrodynamics and sediment transport (Figure 4, red box; hereafter, Grid-f). The initial and lateral boundary conditions for Grid-g were derived from the northwestern Atlantic *ROMS* 50-year solution (courtesy E. Curchitser, Rutgers U.) and the Simple Oceanic Data Assimilation (*SODA*) global reanalysis 50-year dataset (stored as 5-day averages). The annual and monthly temperature and salinity climatology for Grid-g were objectively analyzed from the 1998 World Ocean Atlas. The tidal amplitude and currents (S_2, M_2, K_1, O_1 semidiurnal and diurnal components) forcing for Grid-g's open boundaries were derived from the Oregon State University Tidal Prediction Software (*OTPS*). Both Grid-g and Grid-f included river runoff, and their forcing atmospheric fields were obtained from the European Centre for Medium-Range Weather Forecasts (*ECMWF*) *ERA*-Interim, 3-hour dataset available since 1 January 1978, to the present.

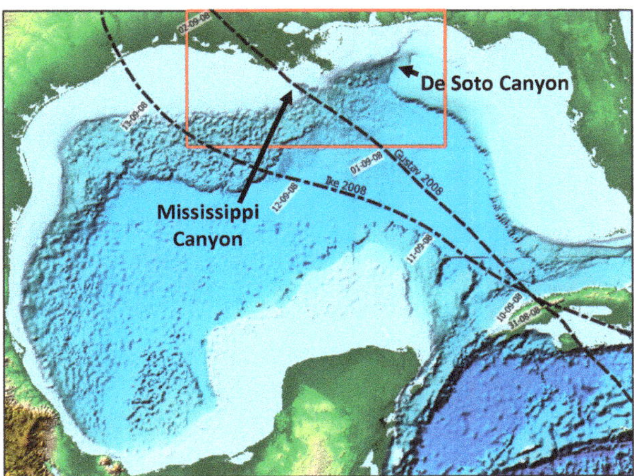

Figure 4. Gulf of Mexico bathymetry showing grid domains from the full coarse grid (Grid-g, black box), and the northern nested grid (Grid-f, red box). Black dashed lines mark approximate tracks of Hurricanes Gustav and Ike. Locations of De Soto and Mississippi Canyons also noted.

The bathymetric grids used by *ROMS* were melded between *ETOPO2* (2 arcminute resolution) and 15arcsecond resolution for the shelf and canyons. Apparent in the bathymetry is the narrowing of the continental shelf near the bird-foot delta, and the presence of both the Mississippi and De Soto Canyons. *ROMS* has terrain-following vertical coordinates, preferred for modeling suspended sediment transport. The bathymetry was smoothed to suppress computational errors in the discretization of horizontal operators (pressure gradient, advection, and diffusion) using a method [42] that allows constraints in the smoothing minimization like preserving the bathymetry in specific grid cells (e.g., on the continental plateau), maximal amplitude modification, desired slope and steepness (**r**-factor), land/sea masking, and preservation of volume.

Spatial and temporal wave data are required to parameterize bottom stress due to wave-current interactions, which affects seafloor sediment transport. *ROMS* requires several wind-induced wavefields to compute bottom stresses from the various bottom boundary layer sub-models available [43]. These fields include significant wave height, wave direction, surface wave period, bottom wave period, bottom orbital velocity, and wave energy dissipation rate. The required fields were processed from the NOAA/NCEP WaveWatch III®dataset (*WW3*; [44]). They were available at three-hour intervals on a grid having a ten arc-minute resolution, and driven by *GFDL-GFS* winds. The *WW3* data were processed from 1 January 2006, to 31 December 2012. Figure 5A,B show wave height and the period during Hurricane Gustav, which impacted the study area during the late summer, 2008. Near-bed wave orbital velocity and near-bed wave periods were estimated from the surface wave characteristics (calculated followng [45]). Figure 5C,D show a sample of bottom wave period and bottom orbital velocity during Hurricane Gustav.

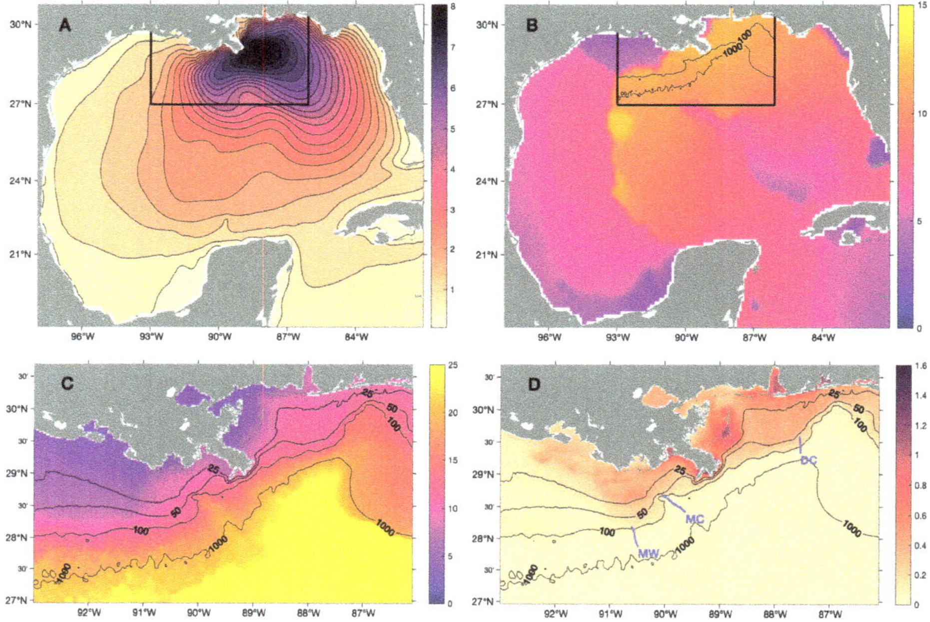

Figure 5. Top row: estimates of wave properties for September 1, 2008 during Hurricane Gustav. (**A**) wave height (m), and (**B**) wave period (s) from NOAA-NCEP *Wavewatch III®* model. Black box indicates the location of *ROMS* hydrodynamic Grid-f, and contours show Grid-f bathymetry (m). Bottom row: calculated estimates of (**C**) bottom wave period (s) and (**D**) bottom orbital velocity (m/s) (calculated following [45]). Transects show locations of flux calculations shown in Figure 9A (MW: Mississippi West); Figure 9B–E (MC: Mississippi Canyon), and Figure 9F–I (DC: De Soto Canyon).

2.2.3. Spatial Seabed Datasets

The *dbSEABED* facility [46–48] supplied information on the spatial distributions of seabed sediment type based on interpolations of more than 10^5 individual data records gleaned from numerous published and unpublished sources. The database provides 0.01-degree resolution mappings of mean grain size (Figure 6A), as well as sorting (*Phi*), gravel, sand and mud fractions (%), exposure of rock (%), and sediment carbonate percent. Local patchiness results from the presence of deep cold-water coral banks [49], low-stand shelf-edge delta remnants [50], and methanogenic carbonate rock and rubble [51]. The shelf areas have important occurrences of gravel, shell, and hard grounds colonized by skeletal-benthos [52].

Sediment stabilities and sediment dispersal patterns are strongly determined by the seabed geomorphology, especially the slope and curvature. To assist the modeling of the generation and then the fate of the turbidity currents, derivative layers were computed from the *SRTM30+* bathymetry, including slope gradients, and location and dimensions of the channelizations. Channel locations are well-discriminated using integration methods such as contributing area; here, the *PyDEM* package [53] was used for such procedure (see background on Figure 6B). On those features, channel-floor dimensions like widths and gradients were computed using operations on the original gridded bathymetry [54], and their values were mainly supplied to the *RANS/TURBINS* turbidity current models.

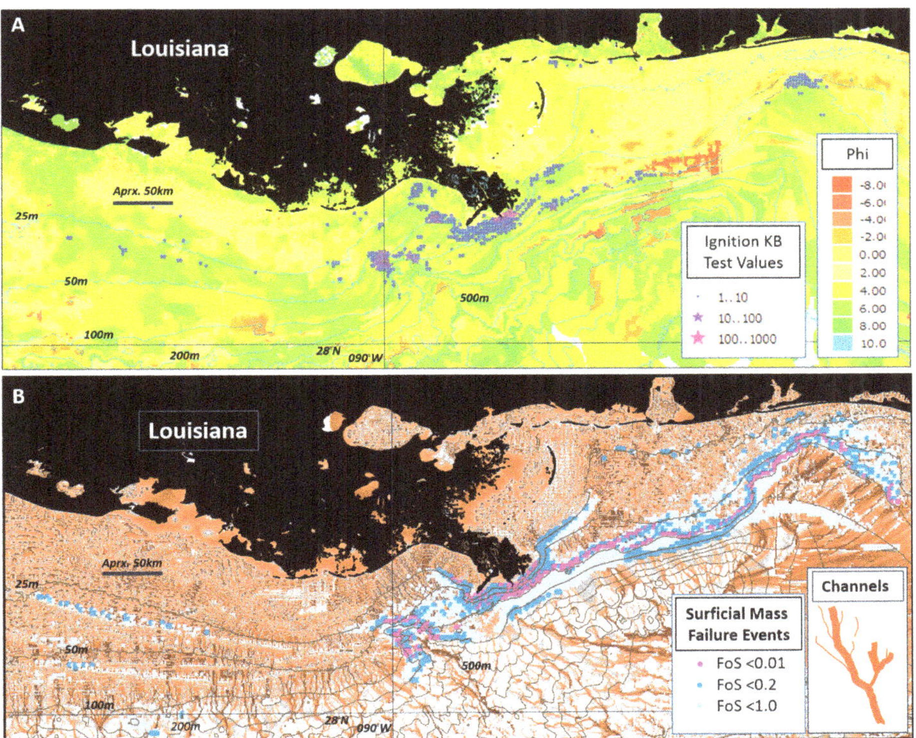

Figure 6. (**A**) The background shows the region's bottom-sediment grain sizes, blue-yellow-red for the range 10 to -8 phi (clay-sand-cobble) from *dbSEABED*. The superimposed points show locations of modeled turbidity current ignitions for Hurricanes Gustav and Ike. Purple points near Mississippi Canyon and Delta mark locations that are particularly prone to ignitions based on density-stability and energy-balance (Knapp–Bagnold, 'KB') measures. The blue points show locations that may also be prone to ignitions. This analysis makes no determination on whether flows will persist over distances. (**B**) The background shows channelized structures (integrated contributing area) for the region, with dark traces marking the most pronounced channels. (Note: The shelf area indications of channeling are dominated by noise in this mapping.) The purple to blue points mark locations of modeled surficial mass failures for Hurricanes Gustav and Ike. The Factors of Safety (*FoS*) indicate the potential for wave-induced mass failure and hence possibly, turbidity currents: purple—very high, blue—high, pale blue—significant. Bathymetric contours shown with depth (m) labeled.

2.2.4. Suspended Sediment Transport Model (*CSTMS*)

The Community Sediment Transport Modeling System (*CSTMS*) has been coupled to the *ROMS* hydrodynamic kernel to represent suspended and bed sediment using user-defined sediment classes; to date, most published *CSTMS* simulations use the non-cohesive routine (see [43]). Each sediment class has attributes of grain diameter, density, settling velocity, and an erosion rate parameter. These are specified in an input file and held constant for the model run. The erodibility of non-cohesive sediment depends on the critical shear stress for erosion (τ_{cr}), specified for each sediment type in an input file. Suspended transport is estimated by assuming that each sediment class acts independently of the others, and travels along with the ambient current velocities, with the addition of the sediment class' settling velocity. The contribution of suspended sediment to water column density is included in the equation of state, and allows for gravitationally driven bottom-boundary layer flows [55,56]. Net exchange of sediment between the seabed and suspended load are estimated by assuming simultaneous erosion, and deposition via settling [43].

We implemented suspended sediment transport simulations for the northern Gulf of Mexico using *CSTMS* on the three-dimensional Grid-f (see Figure 4) from 1 October 2007, through 30 September 2008. This included periods of energetic waves, elevated fluvial discharge, and also Hurricanes Gustav and Ike (Figure 7). Transport and deposition was calculated for seven sediment classes, representing fluvial and seabed sources (Table 1). Fast- and slow-settling sediment was simulated for riverine sediment from the Mississippi, Atchafalaya, and Mobile Rivers. Discharge and sediment concentrations were derived from USGS gauges. Model calculations included estimates of suspended sediment concentration and flux at each location in the three-dimensional *ROMS* Grid-f, and sediment deposition and erosion at each of the horizontal grid points. While the model timestep was 20 s, the output data was saved at hourly intervals.

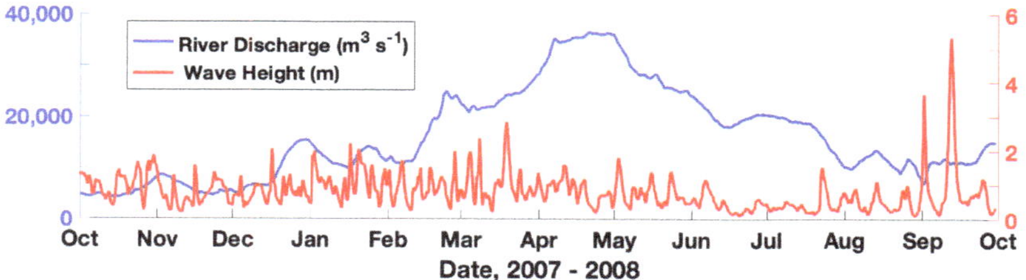

Figure 7. Observed Mississippi River discharge (USGS), and wave height (NOAA's NDBC Buoy #42889) during the modeled period. Hurricanes were in the Gulf of Mexico between 30 August–1 September (Gustav) and 10–13 September (Ike), 2008.

Table 1. Parameters for the suspended sediment transport model. Three sediment classes represented the initial seabed, two sediment classes were discharged by the Mississippi River, and two sediment classes were discharged by the Atchafalaya and Mobile rivers. Critical shear stress and settling velocity for these were based on previous studies [19,20].

Sediment Class	Source	Sediment Type	D (mm)	T_{cr} (Pa)	w_s (mm/s)
1		Mud	0.063	0.11	1.0
2	Seabed	Sand	0.125	0.13	10.0
3		Gravel	10.0	10.0	70.0
4	Mississippi River	Small Mud	0.015	0.11	0.1
5		Larger Mud	0.063	0.11	1.0
6	Atchafalaya/Mobile Rivers	Small Mud	0.015	0.03	0.1
7		Larger Mud	0.063	0.03	1.0

2.2.5. Turbidity Current Ignition Models

A package of one-dimensional, time-dependent, process-numerical modeling modules was used to investigate conditions for wave-induced sediment resuspension and mass wasting, which could potentially lead to turbidity current ignitions. Turbidity currents are known to be generated during events of intense sediment resuspension and mass failure, especially over sloping seafloor [57,58].

The inputs to the modeling package *HurriSlip* included a three-hourly spatially-gridded wave climate based on WaveWatch III®data, surficial seabed material properties from *dbSEABED*, and slope calculations derived from the *SRTM30+* bathymetry. Whereas the *CSTMS* suspension model uses the significant orbital velocity to calculate bed stresses, the implementation of *HurriSlip* relied on a more energetic member of the wave spectra ($H_{1/10}$) to represent resuspension by extreme waves. The predicted sediment failure and ignition events were passed to the *RANS/TURBINS* model-suite, which could simulate the subsequent turbidity current flows down the continental slope. For the predicted cases, the starting flow height, suspended sediment concentration and grain size, and flow velocity were provided. The focus of the work with *HurriSlip* was on the scale of 0-20 m above the seabed with a horizontal resolution of about 1 km.

Sediment resuspension sources: The primary sub-module, *SuspendiSlip*, tested for a likely distribution of turbidity flows arising from wave-induced resuspension of surficial bottom sediment. Most sediment suspension in the continental shelf is thought to be from wave activity during storms [19]. Under significant wave action, bottom-water layers hold significant suspended sediment and turbulent kinetic energy. The module computed several criteria about the ignition of flows. That is, the transformation from bottom waters having significant sediment loading and density to self-sustaining, downslope density-flows undergoing an auto suspension process [59], which allows them to travel for long distances at high speeds. The sediment-laden bottom-water layers were tested from a reference height corresponding to the wave boundary layer thickness up to a height of significant suspersion in a Rouse profile. Sediment pickup was modeled using the excess-over-critical bed shear stress for the sediment, using different formulations for muds [60] and sands [61]. Those published formulations focus on granular erosion at low velocities (mostly <0.5 m/s). However, fine sediments under extreme bed shear during storms are known to erode by bulk-failure [62,63]. The fine sediment erosion rates were capped at the values reported in the publications for the highest bed shear stresses to allow for this.

The bulk, densiometric, Richardson Number (Ri, non-dimensional) divides layers between subcritical (>1.0) and supercritical (<1.0) on the value of $Ri = (g\,R\,C\,h)/U^2$, which depends on gravitational acceleration (g, m/s^2), sediment grain immersed specific gravity (R, non-dimensional), suspended sediment concentration (C, ppm v/v), flow thickness (H, m), and flow velocity (U, m/s). Flows in supercritical disequilibrium are observed to form sustained turbidity currents [64]. The turbulence-supporting flow velocity of layers is reported, based on bottom orbital velocity and ambient currents. For the wave characteristics, Airy linear wave theory was employed. Water properties were not relevant to this calculation; the work of the gravity flow is based on density contrasts due to the suspended sediment.

The primary criterion for ignition was the Knapp–Bagnold criterion (Equation (14)b from [59]), which approximately relates the necessary energy balance $(US)/w_s > 1$, formulated with the seabed gradient (S, non-dimensional) and the grain settling velocity (w_s, m/s). Note that other criteria involving sediment and water entrainment (Equation (16) from [59]) apply to later flow-stages and are less relevant to initial ignition. The modeled events which satisfied the criteria were logged with their associated parameters, and collated onto a mapping (Figure 6A).

Mass failure sources: Large-scale mass failure events are also known to yield or transform into turbidity currents that can travel much further and faster than the original failure structure or debris flow [65,66]. The *WaveSlip* submodule tested for a wave-induced mass failure of seabed sediment during storms based on the circular failure approach [67]. It proved an array of plausible failure arcs, depths, and footprints. Cyclic force moments for each wave period were combined with gravitational moments, and those driving forces were balanced against resisting ones (e.g., gradient shear strength) to test for mass failure. The complex interplay between wave-induced pressures, the footprints of loadings, sub-bottom depths of possible failure arcs, and the gravity-driven and wave-driven moments was integrated into a seafloor Factor-of-Safety (*FoS*) where values less than unity imply instability—the situation of interest.

The possibilities of mass failure were explored for three particular seabed conditions: (i) static undrained conditions of intact shear strength; (ii) remolded shear strengths considering cyclic wave-induced shear strains in the bottom; and (iii) in the presence of liquefaction, especially in shallow waters under long-period surface-waves. For (i), the static shear strengths were calculated using look-up values for the Mississippi Delta area [67]. Remolded values (ii) were computed from those based on wave-induced strains (after [68]). They were scaled linearly against a full remolding to 30% of the intact strengths occurring at 15% cumulative strain. To assess liquefaction potential (iii), a dedicated submodule *LiquiSlip* compared results using previous analytical solutions i.e., [69,70]. Significant wave heights and periods (H_s, m; T_p, s) were assumed to hold for more than 100 wave cycles, and were extracted from *WaveWatch III®* data for each modeled location and time. (Cases of breaking waves were excluded from the analysis; see [70]). Required values for seabed porosity, cohesion, permeability, and relative density (after [71,72]) were calculated based on surface sediment type from *dbSEABED*. The sediment thickness, for which only sparse sub-bottom data exists, was assumed to be effectively infinite. Note that this assumption will not apply in areas <30 m water depth where a "basal, erosional unconformity" at approximately 10 m sub-bottom marks the presence of a firm foundation under Holocene sediments (see [73]). Our study excluded such shallow areas. Time-series of the essential parameters were plotted (not shown) for selected sites in the area to monitor the *WaveSlip* and *LiquiSlip* modeling components.

After the modeling, which took place through the approximately 9 million cell spatial-temporal domain of the project, events at the lowest slope-stability *FoS* were collated and plotted, culminating in the mapping of failure predicted events (Figure 6B). All modeled mass failure events were indicated as potential sites of associated turbidity current ignitions, and their details were passed to the *RANS/TURBINS* component.

2.2.6. *RANS/TURBINS:* a RANS Sediment Gravity Flow Model

TURBINS [74,75] solves the incompressible Navier–Stokes equations in the Boussinesq limit with a convection-diffusion equation for the sediment concentration of small, polydisperse particles whose density significantly exceeds fluid density [76,77]. As a three-dimensional, time-dependent model, *TURBINS* provides spatially and temporally resolved information about the turbulent velocity and sediment concentration fields, conversion of potential into kinetic energy, and the dissipation of this kinetic energy neglecting the effects of rotation. The dispersed phase is assumed to be sufficiently dilute so that the momentum equation governs the two-way coupling between the fluid and particles; the effect of particle loading in the continuity equation is neglected, as are particle interactions such as hindered settling. Particles are assumed to have an aerodynamic response time much smaller than typical fluid flow time scales [78]. Hence, the particle velocity is given by the sum of the fluid velocity and the constant settling velocity. Polydisperse distributions are implemented by considering different particle size classes, each assigned a settling velocity, and contributing to the overall fluid density distribution. Though there is a potential for non-Newtonian dynamics in the dense suspension region near the seafloor, *TURBINS* includes Newtonian fluid dynamics enabling the erosion and resuspension boundary conditions used within the gravity flow module.

An empirical formula to represent the resuspension flux of sediment into the current [79] has been used to estimate erosion in low Reynolds number simulation of turbidity currents [10]. A variation of this was implemented in the non-hydrostatic *RANS/TURBINS* code. The sediment flux due to erosion was introduced into the current as a diffusive flux from the bottom wall.

$$-\frac{1}{Sc\ Re}\frac{\partial c}{\partial \eta} = u_s E_s \tag{1}$$

where c is the non-dimensional concentration of the sediment, η is the coordinate along the direction normal to the boundary, u_s is the settling velocity, E_s is the resuspension flux, Sc is the Schmidt number, and Re is the Reynolds number. Based on [79], the resuspension flux, E_s, was evaluated using

$$E_s = \frac{1}{C_0}\frac{aZ^5}{1 + \frac{a}{0.3}Z^5}; \tag{2}$$

with $a = 1.3 \times 10^{-7}$, C_0 is the initial volume fraction of the sediment, and Z is the erosion parameter. A maximum of $0.3/C_0$ caps the resuspension flux. The erosion parameter, Z, is calculated as

$$Z = 0.586\frac{u_*}{u_s}Re_p^{1.23} \text{ if } Re_p \leq 2.36;$$
$$Z = \frac{u_*}{u_s}Re_p^{0.6} \text{ if } Re_p > 2.36; \tag{3}$$

where u_* is the shear velocity at the bottom wall,

$$\frac{u_t}{u_*} = \frac{1}{\kappa}\log\left(\frac{\eta u_*}{\nu}\right) + B; \tag{4}$$

and u_t is the tangential velocity at the first grid point off the wall, η is the wall-normal distance of the first grid point from the bottom wall, ν is kinematic viscosity, and constants $\kappa = 0.41$ and $B = 5$. Using d_p as the particle diameter, ρ_p as sediment density, ρ_0 as water density, and g as gravitational acceleration; the particle Reynolds number, Re_p, is defined as:

$$Re_p = \frac{d_p\sqrt{gd_p(\rho_p - \rho_0)/\rho_0}}{\nu}. \tag{5}$$

As a proof-of-concept, *TURBINS* was used to represent a turbidity current generated by a lock-release (results in Section 3.3). The lock-release type simulation extended over a 21 km long domain in the streamwise (along the pathway) direction. In the vertical direction, the water depth varied from 130 m to 300 m. Dictated by a minimum resolution criterion of at least ten grid nodes over the current height, along with the condition that the grid spacing is similar in all directions, the simulation employed a grid spacing of 3 m in all directions. Consequently, the computational grid applied 7000 nodes in the streamwise direction, 100 nodes in the vertical direction, and 10 nodes in the spanwise direction. A time step of 0.6 s was used, based on a modified *CFL* (Courant–Friedrichs–Lewy) condition (*CFL* <0.5) involving both convective and viscous terms.

2.3. Modeling Approach

To account for the multi-scale physics of sediment delivery from rivers to the Gulf of Mexico, and subsequent mobilization by oceanic flows, the workflow was designed to operate as follows. Each model component was designed to deliver needed model inputs to the "downstream" models in a one-way coupling framework (Figure 1). Phasing of model development required coordination among the subject matter experts who developed various components of the workflow. The river discharge model (*WBMsed*) can provide values needed as input to the hydrodynamic ocean model. *ROMS* can use these discharges as point sources of freshwater and sediment, distributing the output from *WBMsed* for individual rivers onto the three-dimensional grid of the hydrodynamic model. For example, *WBMsed* provided Mississippi River discharges that were distributed to 39 Mississippi

River discharge grid cells that were spread around the bird-foot delta of the *ROMS* Grid-f. Then, using input winds and open boundary conditions from the lower resolution Gulf model (Grid-g), the local model (Grid-f, see Figure 4) was used in the *CSTMS* to estimate the dispersal and deposition of sediment delivered from the rivers. The bed stresses calculated by *ROMS* accounted for wave-current bed shear stress, and along with *WaveWatch III®* data, could be employed by the *HurriSlip* modules to identify times and locations of sediment mass-failure and density-flow ignition. These events detected by *HurriSlip*, could be used to trigger a turbidity current calculation via *RANS/TURBINS*; which would also be informed by topographic gradients, the sediment properties from *dbSEABED*, and near-bed current velocities and sediment depositions calculated by the *ROMS/CSTMS*.

3. Results

The sections below describe model calculations from components of the workflow to demonstrate their capabilities.

3.1. Suspended Sediment Transport

The *ROMS/CSTMS* ocean model calculated current velocities, bed stresses, suspended sediment fluxes, and erosion/deposition from October 1, 2007, through September 30, 2008. *CSTMS* results for 2007–2008 indicated that the overall signature of sedimentation calculated from suspended sediment was deposition near fluvial sources, with patchy erosion and deposition elsewhere (Figure 8A). Sediment delivered by the Mobile River was largely retained within Mobile Bay. The Atchafalaya River sediment was deposited near the delta, but resuspension events on the inner shelf (depths < 30 m) created westward sediment transport along the coast. Mississippi River plumes more widely dispersed sediment around its bird-foot delta with some of the river load deposited in deeper water (>200 m). The model indicated that the deep sea experienced strong intermittent currents capable of mobilizing sediment, termed benthic storms [80].

Analysis of suspended sediment delivery to the continental slope indicated that about 70% resulted from delivery during low-intensity storms such as frontal systems, and fallout from the Mississippi River plume. The remaining 30% of the year-long delivery of suspended sediment to the continental slope occurred rapidly, during the days surrounding the passage of Hurricanes Gustav and Ike. This supports our first hypothesis, that episodic sediment transport down a submarine canyon is fed by sediment input from wave and current resuspension. Shelf erosion during non-hurricane times accounted for a small fraction of the cumulative erosion seen for the year (Figure 8B). The patchiness of erosion seen in the deep sea (Figure 8B) corresponded to the sediment texture assumed by the model (see Figure 6A). Hurricanes Gustav and Ike created widespread erosion on the shelf, and this material contributed disproportionately to sediment delivery from the shelf to the slope, compared to other resuspension events during the preceding eight months when elevated Mississippi River discharge also occurred (Figure 8). Bed shear stresses during the hurricanes were sufficient to suspend fine-grained sediment across the shelf break. Hurricanes Gustav and Ike produced distinct patterns of erosion and deposition (Figure 8C,D), mainly due to their differences in strength, duration, and storm track (see Figure 4). In general, Ike created higher bed stresses, sediment concentrations, and erosion; but in some locations, Gustav had more impact.

Suspended sediment fluxes along three cross-slope transects (locations shown on Figure 5D) were analyzed to evaluate the phasing and magnitude of hurricane-driven sediment delivery to the continental slope and beyond. The size of sediment flux generally decreased with water depth across the continental slope (Figure 9). While peak fluxes on the continental shelf coincided with the passage of hurricanes, there was often a lag of several days before suspended sediment reached deeper waters. Along the western continental slope, suspended sediment fluxes were larger for Hurricane Ike than Gustav and decreased in the deeper waters (Figure 9A). The peak suspended sediment fluxes at depth (~1300 m) occurred several days after the peak fluxes calculated for the shelf–slope break (~128 m). For the De Soto Canyon, the model estimated net downslope flux towards the south to southeast during the hurricanes (Figure 9F–I). Suspended sediment flux actually increased offshore there between depths of ~650–1115 m during and after the passage of the hurricanes, suggesting that suspended sediment transported over the sides of the canyon settled to the bottom boundary layer and contributed to the suspended sediment fluxes calculated within the canyon (Figure 9F–I).

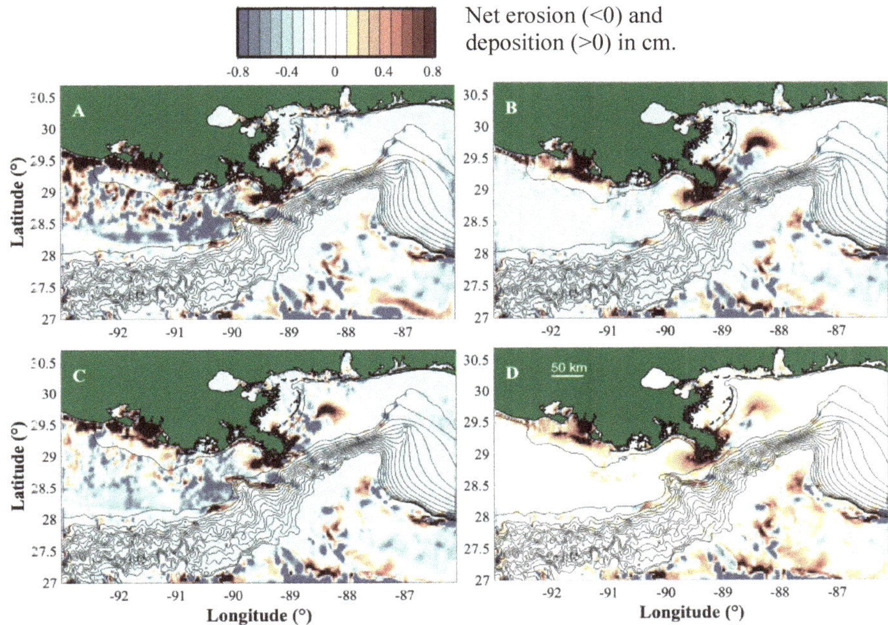

Figure 8. Net erosion (<0) and deposition (>0) calculated for suspended sediment transport for (**A**) entire model run (1 October 2007–20 September 2008); (**B**) prior to hurricanes: time-integrated from 1 October 2007 up until 25 August 2008; (**C**) during Hurricane Gustav; (**D**) during Hurricane Ike. Bathymetric contours (in black) drawn for depths of 10, and every 100 m up to 1500 m.

Sediment fluxes down the Mississippi Canyon lagged behind the passage of the storms, being delayed by 1–5 days relative to the occurrence of peak wave energy on the shelf (Figure 9B–D). The model results showed that these lags corresponded to the time needed for nepheloid layers generated by cross-shelf transport of storm resuspension to be carried to, and settle into, continental slope depths. For example, Hurricane Gustav made landfall on 1 September 2008. For the Mississippi Canyon transect, the model indicated that Gustav created peak sediment fluxes on the outer continental shelf (water depth 98 m) around 2–3 September; while offshore, sediment fluxes did not peak until 4 September (688 m depth) and 8 September (1008 m depth) (Figure 9B,C). The distance along the Mississippi Canyon transect from the 98 m deep site to the 1008 m deep site is about 66 km, so the ~4.5 day lag in delivery to the 1008 m site can be explained by an average horizontal transport velocity of about 0.16 m/s. Similarly, vertical settling delays a storm's impact on deep-sea locations. The fine sediment classes used in the model would settle about 10 or 100 m per day, so that fall out from nepheloid layers would require days to weeks to reach the near-bed continental slope and deeper. This process is illustrated using modeled suspended sediment concentrations along the Mississippi Canyon when Gustav was centered over the Lousiana shelf (Figure 10A), and five days later, the nepheloid layer was delivered to, and settled into, continental slope depths (Figure 10B).

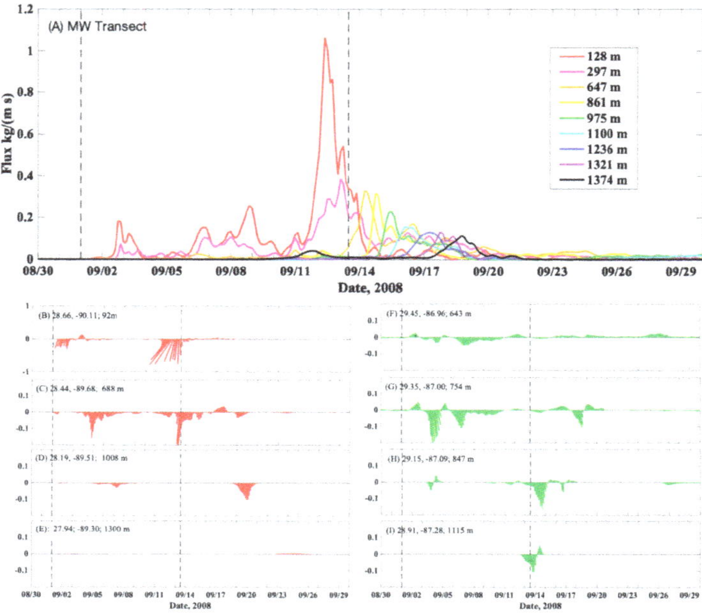

Figure 9. Time-series of depth-integrated suspended sediment flux (kg m^{-1} s^{-1}) calculated along three cross-shelf transects for summer, 2008. Suspended sediment flux across the (**A**) western slope, (**B–E**) Mississippi Canyon (red vectors), and (**F–I**) De Soto Canyon (green vectors). Vector angles correspond to flux direction in accordance with map conventions (down means southward flux). Water depth, latitude, and longitude of each calculation provided as text on figure panels. Locations of transects shown in Figure 5D. Dashed lines mark landfall times of Hurricanes Gustav (9/1) and Ike (9/13).

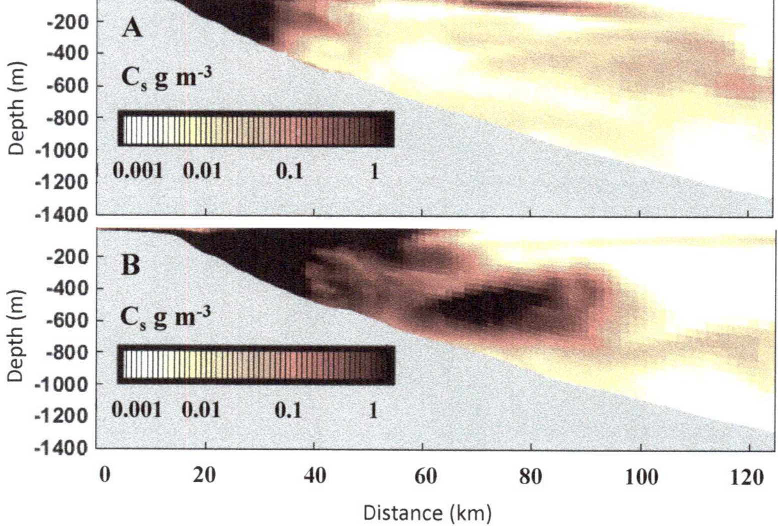

Figure 10. Suspended sediment concentrations calculated along the Mississippi Canyon transect during and after Hurricane Gustav show that sediment delivery to the mid-Canyon lagged several days behind peak storm conditions on the shelf. Model estimates for (**A**) 1 September 2008, and (**B**) 6 September 2008.

3.2. Density Flow Ignitions

Results from the *HurriSlip* model suggest that during extreme storms, bed stresses are large enough to create conditions suitable for the ignition of turbidity currents from near-bottom layers of suspended sediment, especially in areas near the shelf break (Figure 6A). The modeling also suggests that small-thickness sediment mass-failure events, which may evolve into turbidity currents, are widespread around the shelf-slope transition under hurricane conditions (Figure 6B). There is some association between predicted ignitions' locations, and the geomorphic channelizations of the upper continental slope (Figure 6B).

Sediment resuspension sources: The results on the resuspension of sediments into bottom waters (*SuspendiSlip*) indicated suspended sediment concentrations (*SSC*) during times of wave activity averaged ~300 ppm v/v, up to ~5000 ppm v/v (5% v/v) at levels 1 m above the bottom. During the storm events, in shoreface areas including at the delta front, some wave-induced bottom orbital velocities >4 m/s were indicated. At depths of 20–40 m this was reduced to >2 m/s. As modeled, wave-induced resuspensions occurred down to water depths of 189 m (at surface wave periods >13 s) in areas not sheltered from the storm wave effects.

Numerous density-flow ignition events were indicated. They were overwhelmingly in the bottom 1–2 m of the water column, but occasionally occupied water masses as thick as 8 m or more. Bulk densimetric Richardson Number values for the bottom flows ranged widely, but during the storm events were <<1.0 near-bottom i.e., were supercritical states susceptible to the onset of density flow [64]. The Knapp–Bagnold criterion discriminated events more closely and with the gravity influence of slope, identified locations of plausible density flow ignition (Figure 6A). There is some indication that suspension events in the waxing and waning of a storm are more likely to ignite because of the balance between densities and velocities.

Mass failure sources: In agreement with the extensive evidence of mass sediment failures in the region [11,81], the modeling indicated a potential for seafloor failures due to the combined effects of intense storm wave activity, shallow depth, and significant slope. The present prediction with *WaveSlip*, however, also extends over sandy areas not only the mudslide province at the Mississippi Delta front. There seems to be an increased potential for the failure to transform into turbidity current in sandy sediments [82].

Wave-induced liquefaction was predicted in the modeling for conditions of <30 m water depth, somewhat sandy sediments, surface wave wavelengths of >150 m, and significant wave heights of 10 m. Developed (residual and momentary) normal pore pressure increases to exceed normal overburden pressure were modeled down to subbottom depths of 10 m and more at some locations. In those circumstances, effective shear strength was reduced to near zero. The possibility of cyclic strain reduction of shear strengths was also investigated. However, the cumulative strains induced by waves, even during extreme events, were insufficient to produce significantly lowered (remoulded) shear strengths, the strains being at most of order 10^{-2} cumulative (10^{-4} to 10^{-6} per cycle).

The circular-slip analyses indicated mass-failure instabilities (*FoS* <<1.0) over broad areas of sloping seafloor in the top 0.5 m of the seabed (Figure 6B). More deeply-seated failures, down to 20 m sub-bottom, were predicted at a small number of locations at about 30 m water depth. Still, all had a *FoS* >>2 and, therefore, apparently limited potential for actual failure. (They are also at the limits of the analysis in terms of wave-breaking and infinite sediment column.) *HurriSlip* results appear to suggest that without liquefaction or remoulding, probably very few significant wave-induced mass failures would occur in the region. However, the smaller occurrences which are also predicted, remain as candidates to release turbidity flows. They include particularly, many locales with a high likelihood of failure (*FoS* <<1) during storms, in seabed areas down to 100 m water depth, with a significant slope, and often near to the shelf edge.

3.3. Turbidity Currents

As a proof-of-concept for simulating turbidity currents in the northern Gulf of Mexico, we employed *TURBINS* for a lock-release type simulation for a narrow slice along a specific pathway in the failure location. Initially, the normalized sediment concentration, a proxy for excess density due to suspended particles, was set to one in the lock-region and zeroed elsewhere (Figure 11A). When the lock was released, a gravity-flow with a height of 30 m started to travel down the slope, so at 4 h past the lock release, the current was about 10 km downslope (Figure 11B). The current became diluted as a result of entrainment of ambient ocean water, and its height was reduced below the original height of 30 m as it traveled down the slope. Within about 8 h, the current had traveled 15 km down-slope (Figure 11C).

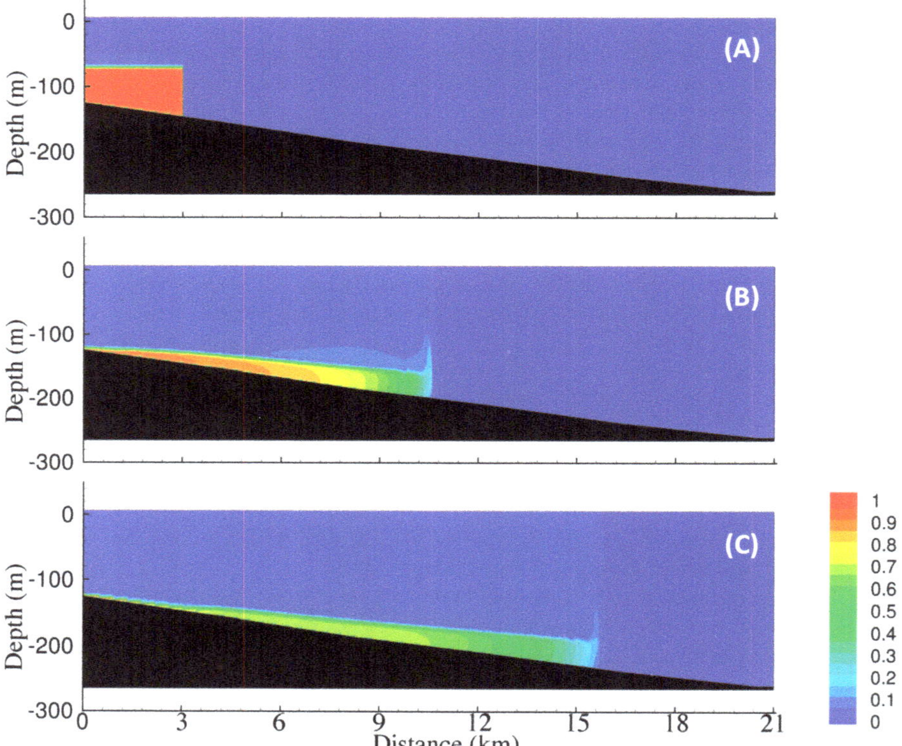

Figure 11. Contours of sediment concentration at (**A**) 0 h, (**B**) 4 h, and (**C**) 8 h. Sediment concentration normalized to values between 0 and 1; 0 indicates clear water without any sediment.

The velocities resulting from the momentum balance are shown in Figure 12 for different stages of the turbidity current. With time, as the turbidity current traveled downslope, thinned and became diluted; its velocities decreased (Figure 12). At 4 h post-ignition, the turbidity current had speeds exceeding 1 m/s, but by 8 h post-release the velocities were much lower. As the current traveled along the bed, it generated a counter-flowing current above that moved in the opposite direction (Figure 12A,B). The calculated velocity at the front of the turbidity current decreased from over 1 m/s to about 0.75 m/s over a period of 10 h (Figure 12C).

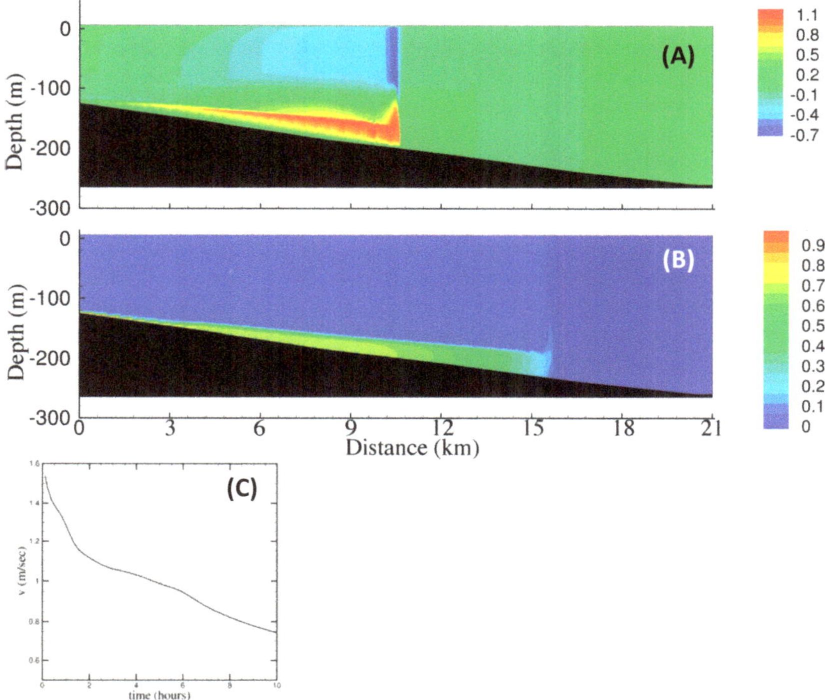

Figure 12. Horizontal velocity (m/s) calculated by TURBINS downslope at (**A**) 4 h, and (**B**) 8 h. Note change in color scale between panels (**A,B**). (**C**) Time history of the front velocity of the current.

Figure 13 displays the modeled bed shear stress at two instants in time. A substantial level of bed shear stress exists along the current length as a result of the turbidity current created by the suspended sediment, which drives the flow. As the current decelerated, the bed shear stress value decreased. These levels of bed stress exceeded the critical shear stress levels for the seabed assumed by the suspended sediment transport model (~0.1 Pa, Table 1), indicating that the gravity flows could be auto suspending, though this process was neglected in this version of the modeling workflow.

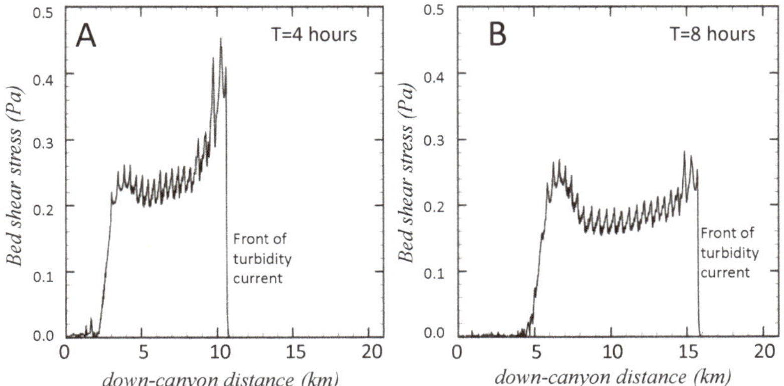

Figure 13. Bed shear stresses (Pa) generated by a simulated down-canyon flowing turbidity current at (**A**) 4 h, and (**B**) 8 h after flow initiation at the shelf-slope break.

4. Discussion

Our results offer estimates of northern Gulf of Mexico sediment delivery and oceanic transport conditions, including locations for gravity flow; and routing of riverine and shelf sediment into submarine canyons. With efforts such as these, that treat multiple time- and space-scales, modeling tools can be developed to deepen our understanding of how sediment is carried from riverine sources to various oceanic sinks. The challenges of integrating various modeling approaches across different spatial and temporal scales are substantial and require further research and code development. Both physical aspects (the implementation of erosion, resuspension of complex sediments into large-scale simulations), as well as numerical challenges (two-way coupling, temporal and spatial interpolation at the boundaries between models) require an additional community effort. The treatment of physical phase-transitions, such as between wave-supported suspended sediment flows and actual turbidity currents, requires more fundamental research.

The models developed for the workflow operated over a broad range of spatial and temporal scales. For example, the *RANS/TURBINS* model represented relatively thin (tens of meters) turbidity currents at higher temporal (<1 s) and spatial (~3 m) resolution than afforded by *ROMS'* hydrodynamic and suspended sediment transport model. As a first step toward multi-scale modeling at the spatial level, our workflow follows sediment routing from the watershed scale via *WBMsed*, to the continental shelf scale via *ROMS*, to specific sediment gravity flows via the *HurriSlip* modules and *RANS/TURBINS*. Likewise, the processes encompassed in our workflow operate over a range of temporal scales, from that of hours for the *TURBINS* model, to the timescale of storm fluctuations for riverine delivery, flow ignition, and suspended transport. Changes in sediment transport that operate at seasonal and interannual timescales are likewise built into our workflow by using forcing functions for weather that represent variations in winds, precipitation, and air temperatures that operate at these timescales. Barriers in applying our methods to longer timescales (i.e., longer than decadal) include both computational limits, and difficulties in assuring that subtle biases in the models and their parameterizations do not cause the calculations to drift from realistic conditions.

The model workflow presented here is sequential, with limited two-way coupling. A fairly straightforward step is to link the riverine discharge model (*WBMsed*) to the oceanic *ROMS* and *CSTMS* models. It would facilitate studies aimed at quantifying oceanic dispersal of fluvial sediments for poorly gauged river systems [83]. Future efforts should explore a more direct model coupling between the suspended sediment transport and gravity flow mechanisms. Within this workflow, *ROMS* estimates the bed shear stresses, which can be used for the flow ignition model (*HurriSlip*). Locations of a slope failure can trigger simulation of a gravity current (e.g., Figure 11), which moves sediment downslope. More direct coupling between these modules would account for sedimentation via suspended sediment transport within the slope failure module, and for net erosion and deposition via gravity currents within the regional scale (*ROMS*) resuspension model.

Regional modeling in the northern Gulf of Mexico is not trivial. Sediment transport modeling requires high-spatial-resolution models to resolve the complex and steep bathymetry. The intense coastal circulation, eddy shedding from the Gulf Loop Current [84], and sporadic strong forcing from storms and hurricanes can affect sediment transport pathways across the continental shelf and slope. Therefore, a telescoping grid approach, from coarse (kilometers) to fine (10s of meters) horizontal scales, is required to obtain viable long-term (1–10 years) and affordable computations. Within our implementation, this was realized by using a low-resolution model for the entire Gulf, telescoping to finer-resolution for the region surrounding the bird-foot delta (Figure 3A). A similar approach has been employed to represent decadal-scale sediment transport in the northern Gulf of Mexico [22].

Joint modeling and field experiments are needed to develop reliable sediment transport models for the Gulf of Mexico continental shelf and slope. Sediment depositional data with which to compare the model calculations are severely lacking, especially at the spatial scales considered here. Recent observational efforts, some motivated by the response to the Deepwater Horizon event, have shown that sediment can be mobilized in deep Gulf of Mexico locations [30,85]. Many of our workflow's sediment transport routines were based on parameterizations for other continental shelf systems, or on laboratory measurements. To improve and gain confidence in the models developed for this workflow requires allied field and modeling studies of sediment processes for the northern Gulf of Mexico continental shelf and slope. Because field sampling during and immediately after storm events is inherently challenging, coupled models that are consistent with observed transport processes and sedimentation are needed to characterize conditions during the extreme events most likely to lead to large sediment fluxes in the deep Gulf of Mexico, and which can damage offshore infrastructure (e.g., [1]).

5. Conclusions

A model-data workflow was developed to numerically represent sediment fluxes from fluvial sources on the inner continental shelf to the continental slope. The workflow is perhaps one of the more complex ever attempted for the problem of routing sediment from coastal sources to deep-sea sinks. The range of components (Figure 1) included: (1) database frameworks for sediment texture and bathymetry of the continental shelf and slope environments; (2) hydrology framework to simulate the discharge of water and sediment for multiple (fifteen) rivers geographically distributed along the northern Gulf of Mexico; (3) an ocean modeling framework that combined output from a spectral wave-action model with ocean circulation simulations, as driven by winds, tides and solar radiation; and tuned to the seafloor environments where bottom boundary layer dynamics can be sufficiently represented including the resuspension, transport and deposition of sediment; (4) a seafloor geotechnical modeling framework able to capture the strengthening and weakening of seafloor deposits, under both ambient ocean conditions, and high intensity, short-lived hurricanes; (5) a gravity flow generator able to determine the location(s) and sediment volume(s) displaced; and (6) a high-resolution CFD model able to simulate the development of a turbidity current, including the bottom shear stresses likely to impact offshore infrastructure. The immersed boundary *RANS* approach, in conjunction with the multiple successive streamwise modules, appears to be well suited to perform the Gulf of Mexico turbidity current simulations over the realistic length and time scales.

The workflow was exercised to explore the conditions that trigger episodes of sediment flux on the continental slope where gas and oil infrastructure exist. Several one-way nested grids from coarse to fine were developed to simulate the hydrodynamic circulation, sediment transport, sediment failure, sediment liquefaction, and turbidity currents in the northern Gulf of Mexico. A full Gulf of Mexico *ROMS* domain was run to provide boundary conditions to a higher-resolution grid that better resolved bathymetric features, river runoff, and sediment transport. Ocean hydrodynamic simulations covered the period from 1 January 2000, to 31 December 2005 (spinup), and from 1 January 2006, to 31 December 2012. It allowed us to characterize sediment transport scenarios during diverse forcing events (river discharge, storms, and multiple hurricanes). We ran focused suspended sediment transport solutions from 1 October 2007, to 30 September 2008, a time period that saw very active tropical storms and major hurricanes crossing the study area.

The suspended sediment model indicated that episodic suspended transport down the Mississippi and De Soto Canyons was fed principally by sediment fluxes generated by wave resuspension on the shelf. During the two hurricanes modeled (Ike and Gustav), suspended sediment fluxes were predominantly seaward in the vicinity of the Mississippi and De Soto Canyons. Peak suspended sediment fluxes coexisted with the occurrence of the highest wave-induced bed stresses on the continental shelf, but showed increasingly long delays relative to this timing with distance down the canyon or continental slope. While hurricane conditions only lasted for two brief episodes during the one-year model run, they accounted for about 30% of the sediment delivered from the continental shelf to the slope. Delivery of sediment directly from settling from the freshwater river plume at the canyon head or over the continental slope provided a more gradual source of sediment delivery for the study period from 1 October 2007, to 30 September 2008. Plume delivery and transport during moderate-intensity frontal passages accounted for 70% of the total sediment delivered to the continental slope during the study period.

The workflow applied a newly developed ignitions model, which was used to explore some particular mechanisms for creating turbidity currents as an additional, and perhaps the major, transportation of sediments to the slope and into channelized features there. Modeling of the flows explored physical constraints on the flow velocities and forces.

On the continental slope, turbidity currents can be triggered by slope failure when storm-driven supply forces accumulation of sediment in deeper water and steeper slopes. These appeared intense enough to both erode sediment along the path of the turbidity current and to damage offshore infrastructure. Modeling efforts in the future should explore more two-way coupling along with workflows such as developed here, and take advantage of observational methods for developing model parameterizations and confirming model estimates.

Author Contributions: Author contributions summarized as: conceptualization, H.G.A., G.A., C.K.H., E.H.M., J.S.; methodology, all; software, H.G.A., J.J.B., S.C., E.W.H.H., C.J.J., T.A.K., S.R.; formal analysis, J.J.B., S.C., T.A.K., C.J.J., S.R.; writing—original draft preparation, C.K.H., J.S.; writing—review and editing, all; visualization, H.G.A., S.C., C.J.J., T.A.K., S.R.; supervision, H.G.A., G.A., C.K.H., E.H.M., J.S.; project administration, G.A.; funding acquisition, H.G.A., C.K.H., E.H.M., J.S. Author contributions can also be summarized by the modeling components on which each contributed: workflow (G.A., J.S.); ROMS (H.G.A.); CSTMS (C.K.H., J.J.B., T.A.K.); WBMsed (J.S., S.C.); HurriSlip (E.W.W.H., C.J.J.); TURBINS (E.H.M., S.R.). All authors have read and agreed to the published version of the manuscript.

Funding: J.S. was supported by U Colorado, CSDMS and BOEM funding, H.A., J.B., C.H., E.H., C.J., S.R.,T.K., and G.A. were supported by BOEM funding. E.M. and S.R. were supported by BOEM and NSF. S.C. was supported by the U Alabama and NSF. J.B. and T.K. were partially supported by VIMS. J.B. was partially supported by the U.S. Geological Survey, Coastal and Marine Hazards and Resources Program.

Acknowledgments: The authors appreciate technical support from D. Robertson (Rutgers U.) and D. Forrest (VIMS). Use of trade, firm or product names is for descriptive purposes only and does not imply endorsement by the U.S. Government. The authors acknowledge William & Mary Research Computing for providing computational resources and technical support that have contributed to the results reported within this paper. The authors appreciate input from reviewers (two anonymous, and Mr. R.C. Mickey, USGS) that helped improve the manuscript. This is contribution number 3925 of the Virginia Institute of Marine Science, William & Mary.

Conflicts of Interest: The authors declare no conflict of interest.

References

1. Bryant, W.L.; Camilli, R.; Fisher, G.B.; Overton, E.B.; Reddy, C.M.; Reible, D.; Swarthout, R.F.; Valentine, D.L. Harnessing a decade of data to inform future decisions: Insights into the ongoing hydrocarbon release at Taylor Energy's Mississippi Canyon Block 20 (MC20) site. *Mar. Pollut. Bull.* **2020**, *155*, 111056. [CrossRef] [PubMed]

2. Stone, G.W.; McBride, R.A. Louisiana barrier islands and their importance in wetland protection: Forecasting shoreline change and subsequent response of wave climate. *J. Coast. Res.* **1998**, *14*, 900–915.

3. Puig, P.; Palanques, A.; Martin, J. Contemporary sediment-transport processes in submarine canyons. *Annu. Rev. Mar. Sci.* **2014**, *6*, 53–77. [CrossRef] [PubMed]

4. Talling, P.J.; Paull, C.K.; Piper, D.J.W. How are subaqueous sediment density flows triggered, what is their internal structure and how does it evolve? Direct observations from monitoring of active flows. *Earth Sci. Rev.* **2013**, *125*, 244–287. [CrossRef]

5. Cacchione, D.A.; Pratson, L.F.; Ogston, A.S. The shaping of continental slopes by internal tides. *Science* **2002**, *296*, 724–727. [CrossRef]

6. Traykovski, P.; Geyer, W.R.; Irish, J.D.; Lynch, J.F. The role of wave-induced density-driven fluid mud flows for cross-shelf transport on the Eel River continental shelf. *Cont. Shelf Res.* **2000**, *20*, 2113–2140. [CrossRef]

7. Hutton, E.W.H.; Syvitski, J.P.M. Advances in the numerical modeling of sediment failure during the development of a continental margin. *Mar. Geol.* **2004**, *203*, 367–380. [CrossRef]

8. Hutton, E.W.H.; Syvitski, J.P.M. Sedflux 2.0: An advanced process-response model that generates three-dimensional stratigraphy. *Comput. Geosci.* **2008**, *34*, 1319–1337. [CrossRef]

9. Parker, G. Conditions for the ignition of catastrophically erosive turbidity currents. *Mar. Geol.* **1982**, *46*, 307–327. [CrossRef]

10. Blanchette, F.; Strauss, M.; Meiburg, E.; Kneller, B.; Glinsky, M.E. High-resolution numerical simulations of resuspending gravity currents: Conditions for self-sustainment. *J. Geophys. Res. Ocean.* **2005**, *110*, 1–15. [CrossRef]

11. Obelcz, J.; Xu, K.; Georgiou, I.Y.; Maloney, J.; Bentley, S.J.; Miner, M.D. Sub-decadal submarine landslides are important drivers of deltaic sediment flux: Insights from the Mississippi River Delta Front. *Geology* **2017**, *45*, 703–706. [CrossRef]

12. Ross, C.B.; Gardner, W.D.; Richardson, M.J.; Asper, V.L. Currents and sediment transport in the Mississippi Canyon and effects of Hurricane Georges. *Cont. Shelf Res.* **2009**, *29*, 1384–1396. [CrossRef]

13. Dail, M.B.; Corbett, D.R.; Walsh, J.P. Assessing the importance of tropical cyclones on continental margin sedimentation in the Mississippi delta region. *Cont. Shelf Res.* **2007**, *27*, 1857–1874. [CrossRef]

14. Sterling, G.H.; Strohbeck, G.E. The failure of the South Pass 70 Platform B in Hurricane Camille. *J. Pet. Technol.* **1975**, *27*, 263–268. [CrossRef]

15. Wijesekera, H.W.; Wang, D.W.; Teague, W.J.; Jarosz, E. High sea-floor stress induced by extreme hurricane waves. *Geophys. Res. Lett.* **2010**, *37*, L11604. [CrossRef]

16. Teague, W.J.; Jarosz, E.; Keen, T.R.; Wang, D.W.; Hulbert, M.S. Bottom scour observed under Hurricane Ivan. *Geophys. Res. Lett.* **2006**, *33*, L07607. [CrossRef]

17. Shay, L.K.; Elsberry, R.L. Near-inertial ocean current response to Hurricane Frederic. *J. Phys. Oceanogr.* **1987**, *17*, 1249–1269. [CrossRef]

18. Brooks, D.A. The wake of Hurricane Allen in the western Gulf of Mexico. *J. Phys. Oceanogr.* **1983**, *13*, 117–129. [CrossRef]

19. Xu, K.; Harris, C.K.; Hetland, R.D.; Kaihatu, J.M. Dispersal of Mississippi and Atchafalaya sediment on the Texas–Louisiana shelf: Model estimates for the year 1993. *Cont. Shelf Res.* **2011**, *31*, 1558–1575. [CrossRef]

20. Xu, K.; Mickey, R.C.; Chen, Q.J.; Harris, C.K.; Hetland, R.D.; Hu, K.; Wang, J. Shelf sediment transport during hurricanes Katrina and Rita. *Comput. Geosci.* **2016**, *90*, 24–39. [CrossRef]

21. Zang, Z.; Xue, Z.G.; Bao, S.; Chen, Q.; Walker, N.D.; Haag, A.S.; Ge, Q.; Yao, Z. Numerical study of sediment dynamics during hurricane Gustav. *Ocean Model.* **2018**, *126*, 29–42. [CrossRef]

22. Zang, Z.; Xue, Z.G.; Xu, K.; Bentley, S.J.; Chen, Q.J.; D'Sa, E.J.; Ge, Q. A two decadal (1993–2012) numerical assessment of sediment dynamics in the northern Gulf of Mexico. *Water* **2019**, *11*, 938. [CrossRef]

23. Meade, R.H.; Moody, J.A. Causes for the decline of suspended-sediment discharge in the Mississippi River system, 1940–2007. *Hydrol. Process.* **2010**, *24*, 35–49. [CrossRef]

24. Wright, L.D.; Nittrouer, C.A. Dispersal of river sediments in coastal seas: Six contrasting cases. *Estuaries* **1995**, *18*, 494–508. [CrossRef]

25. Neill, C.F.; Allison, M.A. Subaqueous deltaic formation on the Atchafalaya Shelf, Louisiana. *Mar. Geol.* **2005**, *214*, 411–430. [CrossRef]

26. Goñi, M.A.; Alleau, Y.; Corbett, R.; Walsh, J.P.; Mallinson, D.; Allison, M.A.; Gordon, E.; Petsch, S.; DellaPenna, T.M. The effects of Hurricanes Katrina and Rita on the seabed of the Louisiana shelf. *Sediment. Rec.* **2007**, *5*, 4–9. [CrossRef]

27. Kineke, G.C.; Higgins, E.E.; Hart, K.; Velasco, D. Fine-sediment transport associated with cold-front passages on the shallow shelf, Gulf of Mexico. *Cont. Shelf Res.* **2006**, *26*, 2073–2091. [CrossRef]

28. Pepper, D.A.; Stone, G.W. Hydrodynamic and sedimentary responses to two contrasting winter storms on the inner shelf of the northern Gulf of Mexico. *Mar. Geol.* **2004**, *210*, 43–62. [CrossRef]

29. Panchang, V.; Jeong, C.K.; Demirbilek, Z. Analyses of extreme wave heights in the Gulf of Mexico for offshore engineering applications. *J. Offshore Mech. Arct. Eng.* **2013**, *135*, 031104. [CrossRef]

30. Diercks, A.-R.; Dike, C.; Asper, V.L.; DiMarco, S.F.; Chanton, J.P.; Passow, U.; Deming, J.W.; Thomsen, L. Scales of seafloor sediment resuspension in the northern Gulf of Mexico. *Elem. Sci. Anthr.* **2018**, *6*, 32. [CrossRef]

31. Cohen, S.; Kettner, A.J.; Syvitski, J.P.M.; Fekete, B.M. *WBMsed*, a distributed global-scale riverine sediment flux model: Model description and validation. *Comput. Geosci.* **2013**, *53*, 80–93. [CrossRef]

32. Cohen, S.; Kettner, A.J.; Syvitski, J.P.M. Global suspended sediment and water discharge dynamics between 1960 and 2010: Continental trends and intra-basin sensitivity. *Glob. Planet. Chang.* **2014**, *115*, 44–58. [CrossRef]

33. Syvitski, J.P.M.; Milliman, J.D. Geology, geography, and humans battle for dominance over the delivery of fluvial sediment to the coastal ocean. *J. Geol.* **2007**, *115*, 1–19. [CrossRef]
34. Morehead, M.D.; Syvitski, J.P.; Hutton, E.W.H.; Peckham, S.D. Modeling the temporal variability in the flux of sediment from ungauged river basins. *Glob. Planet. Chang.* **2003**, *39*, 95–110. [CrossRef]
35. Geological Survey. National Water Information System Data Available on the World Wide Web (USGS Water Data for the Nation); 2016. Available online: http://waterdata.usgs.gov/nwis/ (accessed on 1 February 2020). [CrossRef]
36. Brakenridge, G.R.; Cohen, S.; Kettner, A.J.; De Groeve, T.; Nghiem, S.V.; Syvitski, J.P.M.; Fekete, B.M. Calibration of satellite measurements of river discharge using a global hydrology model. *J. Hydrol.* **2012**, *475*, 123–136. [CrossRef]
37. Haidvogel, D.B.; Arango, H.; Budgell, W.P.; Cornuelle, B.D.; Curchitser, E.; Di Lorenzo, E.; Fennel, K.; Geyer, W.R.; Hermann, A.J.; Lanerolle, L.; et al. Ocean forecasting in terrain-following coordinates: Formulation and skill assessment of the Regional Ocean Modeling System. *J. Comput. Phys.* **2008**, *227*, 3595–3624. [CrossRef]
38. Shchepetkin, A.F.; McWilliams, J.C. The regional oceanic modeling system (ROMS): A split-explicit, free-surface, topography-following-coordinate oceanic model. *Ocean Model.* **2005**, *9*, 347–404. [CrossRef]
39. Shchepetkin, A.F.; McWilliams, J.C. A method for computing horizontal pressure-gradient force in an oceanic model with a nonaligned vertical coordinate. *J. Geophys. Res. Ocean.* **2003**, *108*, 3090. [CrossRef]
40. Durski, S.M.; Glenn, S.M.; Haidvogel, D.B. Vertical mixing schemes in the coastal ocean: Comparison of the level 2.5 Mellor-Yamada scheme with an enhanced version of the K profile parameterization. *J. Geophys. Res. Ocean.* **2004**, *109*, C01015. [CrossRef]
41. Warner, J.C.; Sherwood, C.R.; Arango, H.G.; Signell, R.P. Performance of four turbulence closure models implemented using a generic length scale method. *Ocean Model.* **2005**, *8*, 81–113. [CrossRef]
42. Sikirić, M.D.; Janekovic, I.; Kuzmić, M. A new approach to bathymetry smoothing in sigma-coordinate ocean models. *Ocean Model.* **2009**, *29*, 128–136. [CrossRef]
43. Warner, J.C.; Sherwood, C.R.; Signell, R.P.; Harris, C.K.; Arango, H.G. Development of a three-dimensional, regional, coupled wave, current, and sediment-transport model. *Comput. Geosci.* **2008**, *34*, 1284–1306. [CrossRef]
44. Tolman, H.L.; Chalikov, D.V. Development of a third-generation ocean wave model at NOAA-NMC. In *Physical and Numerical Modelling*; Isaacson, M., Quick, M.C., Eds.; The University of British Columbia: Vancouver, BC, Canada, 1994; pp. 724–733.
45. Wiberg, P.L.; Sherwood, C.R. Calculating wave-generated bottom orbital velocities from surface-wave parameters. *Comput. Geosci.* **2008**, *34*, 1243–1262. [CrossRef]
46. Goff, J.A.; Jenkins, C.J.; Williams, S.J. Seabed mapping and characterization of sediment variability using the usSEABED data base. *Cont. Shelf Res.* **2008**, *28*, 614–633. [CrossRef]
47. Jenkins, C. Dominant bottom types and habitats. In *Gulf of Mexico Data Atlas. Stennis Space Center (MS): National Coastal Data Development Center*; 2011. Available online: http://gulfatlas.noaa.gov/ (accessed on 27 June 2020).
48. Williams, S.J.; Flocks, J.; Jenkins, C.; Khalil, S.; Moya, J. Offshore sediment character and sand resource assessment of the northern Gulf of Mexico, Florida to Texas. *J. Coast. Res.* **2012**, *60*, 30–44. [CrossRef]
49. Brooke, S.; Schroeder, W.W. State of deep coral ecosystems in the Gulf of Mexico region, Texas to the Florida Straits. In *The State of Deep Coral Ecosystems of the United States*; Lumsde, S.E., Hourigan, T.F., Bruckner, A.W., Dorr, G., Eds.; NOAA Technical Memorandum CRCP-3; U.S. Department of Commerce: Silver Spring, MD, USA, 2007; pp. 271–306.
50. Gardner, J.V.; Calder, B.R.; Clarke, J.E.H.; Mayer, L.A.; Elston, G.; Rzhanov, Y. Drowned shelf-edge deltas, barrier islands and related features along the outer continental shelf north of the head of De Soto Canyon, NE Gulf of Mexico. *Geomorphology* **2007**, *89*, 370–390. [CrossRef]
51. Cordes, E.E.; Bergquist, D.C.; Fisher, C.R. Macro-ecology of Gulf of Mexico cold seeps. *Annu. Rev. Mar. Sci.* **2009**, *1*, 143–168. [CrossRef]
52. Thompson, D.M.; Wohl, E.E.; Jarrett, R.D. Velocity reversals and sediment sorting in pools and riffles controlled by channel constrictions. *Geomorphology* **1999**, *27*, 229–241. [CrossRef]

53. Ueckermann, M.P.; Chambers, R.D.; Brooks, C.A.; Audette, W.E.; Bieszczad, J. pyDEM: Global digital elevation model analysis. In Proceedings of the 14th Python in Science Conference (SciPy 2015), Austin, TX, USA, 6–12 July 2015; Volume 6, pp. 117–124.

54. Jenkins, C. Sediment Drainage Streams Important in Benthic Seafloor Classification. In *Ocean Solutions, Earth Solutions*, 2nd ed.; Wright, D., Ed.; ESRI Press: Redlands, CA, USA, 2016; pp. 431–440. [CrossRef]

55. Bever, A.J.; Harris, C.K. Storm and fair-weather driven sediment-transport within Poverty Bay, New Zealand, evaluated using coupled numerical models. *Cont. Shelf Res.* **2014**, *86*, 34–51. [CrossRef]

56. Chen, S.-N.; Geyer, W.R.; Hsu, T.-J. A numerical investigation of the dynamics and structure of hyperpycnal river plumes on sloping continental shelves. *J. Geophys. Res. Ocean.* **2013**, *118*, 2702–2718. [CrossRef]

57. Fukushima, Y.; Parker, G.; Pantin, H.M. Prediction of ignitive turbidity currents in Scripps Submarine Canyon. *Mar. Geol.* **1985**, *67*, 55–81. [CrossRef]

58. Piper, D.J.W.; Cochonat, P.; Morrison, M.L. The sequence of events around the epicentre of the 1929 Grand Banks earthquake: Initiation of debris flows and turbidity current inferred from sidescan sonar. *Sedimentology* **1999**, *46*, 79–97. [CrossRef]

59. Parker, G.; Fukushima, Y.; Pantin, H.M. Self-accelerating turbidity currents. *J. Fluid Mech.* **1986**, *171*, 145–181. [CrossRef]

60. Parchure, T.M.; Mehta, A.J. Erosion of soft cohesive sediment deposits. *J. Hydraul. Eng.* **1985**, *111*, 1308–1326. [CrossRef]

61. Nielsen, P. *Coastal Bottom Boundary Layers and Sediment Transport*; Advanced Series on Ocean Engineering; World Scientific: Hackensack, NJ, USA, 1992; Volume 4, 340p.

62. Amos, C.L.; Daborn, G.R.; Christian, H.A.; Atkinson, A.; Robertson, A. In situ erosion measurements on fine-grained sediments from the Bay of Fundy. *Mar. Geol.* **1992**, *108*, 175–196. [CrossRef]

63. Schieber, J.; Southard, J.B.; Schimmelmann, A. Lenticular shale fabricsresulting from intermittent erosion of water-rich muds—Interpreting the rock record in the light of recent flume experiments. *J. Sediment. Res.* **2010**, *80*, 119–128. [CrossRef]

64. Pantin, H.M.; Franklin, M.C. Predicting autosuspension in steady turbidity flow: Ignition points and their relation to Richardson numbers. *J. Sediment. Res.* **2009**, *79*, 862–871. [CrossRef]

65. Mohrig, D.; Marr, J.G. Constraining the efficiency of turbidity current generation from submarine debris flows and slides using laboratory experiments. *Mar. Pet. Geol.* **2003**, *20*, 883–899. [CrossRef]

66. Waltham, D. Flow Transformations in particulate gravity currents. *J. Sediment. Res.* **2004**, *74*, 129–134. [CrossRef]

67. Nodine, M.C.; Cheon, J.Y.; Wright, S.G.; Gilbert, R.B. Mudslides during Hurricane Ivan and an assessment of the potential for future mudslides in the Gulf of Mexico. In *C175 Phase I Project Report*; Minerals Management Service: College Station, TX, USA, 2006; MMS Project Number 552; 37p.

68. Nataraja, M.S.; Gill, H.S. Ocean wave-induced liquefaction analysis. *J. Geotech. Eng.* **1983**, *109*, 573–590. [CrossRef]

69. Geremew, A.M. Pore-water pressure development caused by wave-induced cyclic loading in deep porous formation. *Int. J. Geomech.* **2013**, *13*, 65–68. [CrossRef]

70. Jeng, D.-S.; Seymour, B.; Li, J. A new approximation for pore pressure accumulation in marine sediment due to water waves. *Int. J. Numer. Anal. Methods Geomech.* **2006**, *31*, 53–69. [CrossRef]

71. Boudreau, B.P.; Bennett, R.H. New rheological and porosity equations for steady-state compaction. *Am. J. Sci.* **1999**, *299*, 517–528. [CrossRef]

72. Youd, T. Factors Controlling maximum and minimum densities of sands. In *Evaluation of Relative Density and Its Role in Geotechnical Projects Involving Cohesionless Soils*; Selig, E., Ladd, R., Eds.; ASTM International: West Conshohocken, PA, USA, 2009; pp. 98–112. [CrossRef]

73. Brooks, G.R.; Kindinger, J.L.; Penland, S.; Williams, S.J.; McBride, R.A. East Louisiana continental shelf sediments: A product of delta reworking. *J. Coast. Res.* **1995**, *11*, 1026–1036.

74. Nasr-Azadani, M.M.; Meiburg, E. *TURBINS*: An immersed boundary, Navier–Stokes code for the simulation of gravity and turbidity currents interacting with complex topographies. *Comput. Fluids* **2011**, *45*, 14–28. [CrossRef]

75. Nasr-Azadani, M.M.; Meiburg, E. Turbidity currents interacting with three-dimensional seafloor topography. *J. Fluid Mech.* **2014**, *745*, 409–443. [CrossRef]

76. Necker, F.; Härtel, C.; Kleiser, L.; Meiburg, E. High-resolution simulations of particle-driven gravity currents. *Int. J. Multiph. Flow* **2002**, *28*, 279–300. [CrossRef]

77. Necker, F.; Härtel, C.; Kleiser, L.; Meiburg, E. Mixing and dissipation in particle-driven gravity currents. *J. Fluid Mech.* **2005**, *545*, 339–372. [CrossRef]

78. Maxey, M.R.; Riley, J.J. Equation of motion for a small rigid sphere in a nonuniform flow. *Phys. Fluids* **1983**, *26*, 883–889. [CrossRef]

79. Garcia, M.H.; Parker, G. Experiments on the entrainment of sediment into suspension by a dense bottom current. *J. Geophys. Res. Ocean.* **1993**, *98*, 4793–4807. [CrossRef]

80. Gardner, W.D.; Tucholke, B.E.; Richardson, M.J.; Biscaye, P.E. Benthic storms, nepheloid layers, and linkage with upper ocean dynamics in the western North Atlantic. *Mar. Geol.* **2017**, *385*, 304–327. [CrossRef]

81. Prior, D.B.; Coleman, J.B.; Garrison, L.E. Digitally acquired undistorted side scan sonar images of submarine landslides, Mississippi Delta. *Geology* **1979**, *7*, 423–425. [CrossRef]

82. Elverhøi, A.; Breien, H.; De Blasio, F.V.; Harbitz, C.B.; Pagliardi, M. Submarine landslides and the importance of the initial sediment composition for run-out length and final deposit. *Ocean Dyn.* **2010**, *60*, 1027–1046. [CrossRef]

83. Kuehl, S.A.; Alexander, C.R.; Blair, N.E.; Harris, C.K.; Marsaglia, K.M.; Ogston, A.S.; Orpin, A.R.; Roering, J.J.; Bever, A.J.; Bilderback, E.L.; et al. A source-to-sink perspective of the Waipaoa River margin. *Earth Sci. Rev.* **2016**, *153*, 301–334. [CrossRef]

84. Sturges, W.; Leben, R. Frequency of ring separations from the loop current in the Gulf of Mexico: A revised estimate. *J. Phys. Oceanogr.* **2000**, *30*, 1814–1819. [CrossRef]

85. Larson, R.A.; Brooks, G.R.; Schwing, P.T.; Holmes, C.W.; Carter, S.R.; Hollander, D.J. High-resolution investigation of event driven sedimentation: Northeastern Gulf of Mexico. *Anthropocene* **2018**, *24*, 40–50. [CrossRef]

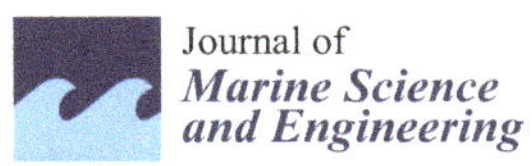

Journal of
Marine Science and Engineering

Article

Evaluation of Structured and Unstructured Models for Application in Operational Ocean Forecasting in Nearshore Waters

Shannon Nudds [1,*], Youyu Lu [1], Simon Higginson [1], Susan P. Haigh [2], Jean-Philippe Paquin [3], Mitchell O'Flaherty-Sproul [2], Stephanne Taylor [1], Hauke Blanken [4], Guillaume Marcotte [5], Gregory C. Smith [3], Natacha B. Bernier [3], Phillip MacAulay [6], Yongsheng Wu [1], Li Zhai [1], Xianmin Hu [1], Jérôme Chanut [7], Michael Dunphy [4], Frédéric Dupont [8], David Greenberg [1], Fraser J. M. Davidson [9] and Fred Page [2]

1 Ocean and Ecosystem Science Division, Fisheries and Oceans Canada, Bedford Institute of Oceanography, 1 Challenger Drive, Dartmouth, NS B2Y 4A2, Canada; Youyu.Lu@dfo-mpo.gc.ca (Y.L.); Simon.Higginson@dfo-mpo.gc.ca (S.H.); Stephanne.Taylor@dfo-mpo.gc.ca (S.T.); Yongsheng.Wu@dfo-mpo.gc.ca (Y.W.); Li.Zhai@dfo-mpo.gc.ca (L.Z.); Xianmin@ualberta.ca (X.H.); David.Greenberg@dfo-mpo.gc.ca (D.G.)
2 Coastal Ecosystem Sciences Division, Fisheries and Oceans Canada, St. Andrews Biological Station, 125 Marine Science Drive, St. Andrews, NB E5B 2L9, Canada; Susan.Haigh@dfo-mpo.gc.ca (S.P.H.); Mitchell.O'Flaherty-Sproul@dfo-mpo.gc.ca (M.O.-S.); Fred.Page@dfo-mpo.gc.ca (F.P.)
3 Meteorological Research Division, Environment and Climate Change Canada, 2121 Transcanada Highway, Dorval, QC H9P 1J3, Canada; Jean-Philippe.Paquin@canada.ca (J.-P.P.); Gregory.Smith2@canada.ca (G.C.S.); Natacha.Bernier@canada.ca (N.B.B.)
4 Ocean Sciences Division, Fisheries and Ocean Canada, Institute of Ocean Science, 9860 West Saanich Rd., Sidney, BC V8L 4B2, Canada; Hauke.Blanken@dfo-mpo.gc.ca (H.B.); Michael.Dunphy@dfo-mpo.gc.ca (M.D.)
5 Canadian Centre for Meteorological and Environmental Prediction, National Operation Division, Environmental Emergencies Section, Environment and Climate Change Canada, 2121 Transcanada Highway, Dorval, QC H9P 1J3, Canada; Guillaume.Marcotte@canada.ca
6 Canadian Hydrographic Service, Fisheries and Ocean Canada, Bedford Institute of Oceanography, 1 Challenger Drive, Dartmouth, NS B2Y 4A2, Canada; Phillip.MacAulay@dfo-mpo.gc.ca
7 Mercator-Océan, International, 10 rue Hermès, 31520 Ramonville-Saint-Agne, France; jchanut@mercator-ocean.fr
8 Meteorological Service of Canada, Environment and Climate Change Canada, 2121 Transcanada Highway, Dorval, QC H9P 1J3, Canada; Frederic.Dupont@canada.ca
9 Ocean and Ecosystem Science Division, Fisheries and Oceans Canada, North Atlantic Fisheries Centre, 80 E. White Hills Rd, St. John's, NL A1A 5J7, Canada; Fraser.Davidson@dfo-mpo.gc.ca
* Correspondence: Shannon.Nudds@dfo-mpo.gc.ca

Received: 30 April 2020; Accepted: 26 June 2020; Published: 30 June 2020

Abstract: The oceanography sub-initiative of Canada's Oceans Protection Plan was tasked to develop high-resolution nearshore ocean models for enhanced marine safety and emergency response, fitting into the multi-scale, multi-level nested operational ocean forecasting systems. For decision making on eventual 24/7 operational support, two ocean models (a structured grid model, NEMO (Nucleus for European Modelling of the Ocean); and an unstructured grid model, FVCOM (Finite Volume Coastal Ocean Model), were evaluated. The evaluation process includes the selection of the study area, the requirements for model setup, and the evaluation metrics. The chosen study area, Saint John Harbour in the Bay of Fundy, features strong tides, significant river runoff and a narrow tidal-river channel. Both models were configured with the same sources of bathymetry and forcing data. FVCOM achieved 50–100 m horizontal resolution in the inner harbour and included wetting/drying. NEMO achieved 100 m resolution in the harbour with a three-level one-way nesting configuration. Statistical metrics showed that one-year simulations with both models achieved comparable accuracies against the observed tidal and non-tidal water levels and currents, temperature and salinity, and the trajectories of surface drifters, but the computational cost of FVCOM was significantly less than that of NEMO.

Keywords: coastal modelling; operational forecasting; model evaluation; inter-comparison; NEMO; FVCOM; Ocean Protection Plan

---

1. Introduction

In 2016, Canada launched the CAD 1.5 billion Ocean Protection Plan (OPP) [1] to protect the world's longest coastline and support cleaner, healthier and safer waters. Under the oceanography sub-initiative of the OPP, Fisheries and Oceans Canada (DFO) was tasked to develop high-resolution operational nearshore ocean models for enhanced marine safety and emergency response, specifically electronic navigation and the prediction of oil spill drift trajectory. The nearshore models will eventually fit into the multi-scale, multi-level nested operational ocean-forecasting systems of the Government of Canada, through the collaborative development by Environment and Climate Change Canada (ECCC) and DFO under the Canadian Operational Network of Coupled Environmental Prediction Systems (CONCEPTS) [2] Memorandum of Understanding. The current phase of this OPP sub-initiative focuses on six pilot ports/waterways: Kitimat, Port Metro-Vancouver, Fraser River Port, the St. Lawrence River, Port Hawkesbury, and the Port of Saint John, with plans to extend modelling to other ports in the future. In 2017, during the first year of OPP, a significant effort was made to develop configurations for the Port of Saint John using two widely used, open source, ocean models and evaluate their suitability for OPP applications. The two models, NEMO (Nucleus for European Modelling of the Ocean) [3,4] and FVCOM (Finite Volume Coastal Ocean Model), [5,6] were selected due to their existing applications in Canada.

NEMO is a finite difference model that runs on structured horizontal grids. It was first developed for global and basin-scale applications, and subsequently for coastal applications. Prior to this study, NEMO had not been used for near-shore, port-scale applications. Under CONCEPTS, DFO and ECCC developed a series of operational ocean and sea-ice prediction systems using NEMO, covering the global ocean (with a horizontal resolution of 1/4-degree in longitude/latitude, [7]), regional ocean basins (North Atlantic, Arctic and North Pacific, with 1/12-degree resolution, [8]), and the Great Lakes (with 2 km resolution; [9]). The development of prediction systems for the shelf and coastal oceans off the western and eastern coasts of Canada (with 1/36-degree resolution) is ongoing. These systems run operationally with 24/7 support at ECCC's Canadian Centre for Meteorological and Environmental Prediction (CCMEP).

FVCOM uses finite volume numerics on unstructured horizontal grids. In Canada, and particularly within DFO, there are extensive applications of FVCOM for coastal, near-shore and lake waters. FVCOM is used for simulating both barotropic tides (without including density variations) (e.g., [10,11]), and full baroclinic dynamics (e.g., [12–15]). The use of unstructured grids enables very high horizontal resolution, reaching a couple of metres in nearshore waters in some cases (e.g., [16,17]) while maintaining coarser resolution for offshore areas with larger scale dynamics. Many applications make use of FVCOM's wetting/drying scheme to simulate processes in the intertidal zone (e.g., [17,18]). Under a previous DFO project, FVCOM was used to develop port scale models for five of the six OPP pilot ports, including a baroclinic configuration for the Port of Saint John (without atmospheric forcing).

Several studies have compared different configurations of the same root model, but few have compared fully baroclinic, structured and unstructured models with similar resolution over the same domain. Huang et al. [19] compared FVCOM with a structured grid model, ROMS (the Regional Ocean Modelling System), [20] but the study focused on idealized test cases and used barotropic configurations only. Trotta et al. [21] examined the use of NEMO and the Shallow Water Hydrodynamic Finite Element Model (SHYFEM) [22], in a downscaling context for a relocatable ocean platform for forecasting, but they did not directly compare the two models. More recently, and most relevant to our study, Biastoch et al. [23] performed a comprehensive comparison of a nested configuration of

NEMO and the unstructured Finite Element Sea Ice-Ocean Model (FESOM) [24], but this was for global configurations and focused on large scale ocean circulation.

The models developed for OPP applications will eventually be run with 24/7 operational support, but amongst the Canadian government agencies, the capacity to do so only exists within the ECCC. ECCC currently uses NEMO for operational ocean forecasting. To add the operation of FVCOM to ECCC would entail additional resources (and cost) compared to utilizing the existing NEMO infrastructure. Hence, ECCC required the assessment of the performance of both NEMO and FVCOM for consideration in decision making. To achieve this, DFO and ECCC jointly developed an evaluation process, summarized here, to objectively compare both the predictive accuracy of the key parameters required for OPP applications, and the efficiency in terms of computational cost, between NEMO and FVCOM.

This paper describes the principles and factors that were considered in designing the evaluation process (Section 2), and application to the Port of Saint John including the metrics and sample results of the evaluation (Section 3). The selected examples do not cover the full results of the NEMO/FVCOM evaluation, and do not demonstrate the full strength of either model. The evaluation guided on-going research in the development and improvement of both models. More comprehensive descriptions of the configurations and results of both models are or will be documented elsewhere, e.g., the NEMO configuration by Paquin et al. [25]. Finally, the proposed process and metrics can be generalized and modified to evaluate the configurations developed for other regions, and with models other than NEMO and FVCOM, for research and operations.

2. Factors Considered for Evaluation

The evaluation process includes the selection of the study area, the requirements for the model setup, and the metrics for evaluating the models. The evaluation was designed to objectively (quantitatively) assess the accuracy and efficiency of NEMO and FVCOM for operational forecasting at port scales for parameters of interest to the OPP applications. Here, the term *accuracy* pertains to the models' ability to reproduce the observations, and *efficiency* refers to the computer resources that are required to run the models in an operational context. The objectives of OPP are to improve electronic navigation and to predict oil spill drift, thus the parameters of interest for the evaluation were water level, currents, water temperature and salinity (density), and surface drifter trajectories. As with any experimental design, it is important to consider the ability of the evaluation to detect contrasts in the models and to force the strongest contrast possible. Thus, the evaluation was designed to challenge the models with respect to accuracy and efficiency. Consequently, the evaluation helped gain a better understanding of the strengths and weaknesses of the models, and although the evaluation focused on one port, the results can be reasonably expected to extend to other ports.

Of the six OPP pilot ports, one was selected for the evaluation process. The selection of the study area was based on (1) the regional oceanography being sufficiently complex to include key dynamic processes; (2) the availability of forcing data (from atmosphere, rivers, and open ocean) to drive the port models; and (3) the availability of sufficient observational data to assess the predictive accuracy of the models with respect to the parameters of interest. The parameters of interest at Canadian ports are typically influenced by complicated coastlines, bathymetry, tides, river runoff, the open ocean, and, in some cases, the presence of sea-ice. Due to the urgent timeline of the OPP (i.e., the evaluation was to be completed within the first year of OPP), the explicit inclusion of sea-ice and the simulation of the intertidal zone (with a wetting/drying scheme) were not required, but to ensure that FVCOM was used to its full potential, the use of the wetting/drying scheme was encouraged. At the time of the evaluation, a wetting/drying scheme was not available in NEMO. Forecasting the surface waves at port scales was not considered for the current phase of the OPP.

Various aspects of the model setup were built into the process of evaluation, including spatial resolution, the inclusion of dynamic processes, forcing fields, the duration of the simulation, computational cost, and the variables, frequencies and format of the model output. The required horizontal grid spacing of the models was 100 m or less to resolve the horizontal gradients of currents and the presence of eddies due to nonlinear processes. The required vertical grid spacing was 1 m or less near the surface, to resolve the currents in the upper layer that are important for navigation and oil spill drift. The models included the full baroclinic dynamics to simulate the variations due to surface momentum and buoyancy fluxes, river runoff, and open ocean forcing into the port. The models were subjected to the same forcing, which included atmospheric forcing from the operational weather forecasting system, large-scale oceanic forcing from the operational regional ice-ocean forecasting system, and available tidal and river runoff. At the current stage of the OPP, the port models did not include any data assimilation, partly due to the lack of sufficient real time observational data. A common time frame for the simulation was determined by the available observations. The duration of the simulation was 15 months to allow for a spin-up of the models and the evaluation of a full annual cycle. For the proper evaluation of model efficiency, both models were run on the same high-performance computer facility of the Government of Canada. For operational applications, the required run-time of the models was 0.5 h (or less) for a 48 h simulation. The frequency of the model output was minimally 0.5 h for the proper evaluation of tidal variations.

The metrics for the model data comparison were defined based on the existing expertise of the team, and through expert consulting and literature research. The metrics were defined for the tidal and non-tidal components of sea level and currents, vertical profiles of water temperature and salinity, time series of sea surface temperature (SST) at fixed stations, and the trajectories of surface drift. The quantitative metrics are statistically robust to measure the discrepancy between the model solution and observational data. Prior to performing the statistical comparison, the model results were extracted from the grid node nearest to the locations of the observations. Time series analysis also included a comparison of the energy spectrum. Because the domains of both models cover the area beyond the port, the evaluation was carried out for the "inner harbour" (port) and "outer harbour" separately, with an emphasis put on the inner harbour. In addition to the quantitative evaluation, the models were evaluated qualitatively based on known features of the regional oceanography, e.g., the presence of the river plume, tidal fronts, eddies, etc.

Finally, when possible, the models were compared to existing operational products that are used in Canada. This includes the Scotia-Fundy-Maine WebTide (hereafter referred to as WebTide) solutions for tidal elevation and currents [26], and the regional ice-ocean prediction system (RIOPS) that covers the North Atlantic, Arctic and North Pacific with 1/12-degree resolution [8]. Note that neither WebTide or RIOPS were developed for near-shore operational applications and are not high-resolution port solutions, but were the existing operational products for the area at the time of this evaluation.

3. Results

This section describes how the evaluation process, outlined in Section 2, was applied to the NEMO and FVCOM models for the Port of Saint John. Samples of the evaluation results are based on the model solutions near the end of the evaluation process (December 2017). The evaluation identified some deficiencies or errors in both models, and the subsequent work beyond this evaluation process led to improved model solutions, e.g., documented in Paquin et al. for NEMO [25].

3.1. Regional Oceanography

The Port of Saint John was selected as the study site, in part because of its complicated regional oceanography. The port is in Saint John Harbour (hereafter SJH) located in the Bay of Fundy (Figure 1). The area presents tides among the largest in the world, with a maximum range reaching 16 m in the upper Bay of Fundy [27]. The tidal range in SJH can exceed 8 m at spring tide [28,29]. SJH receives significant freshwater influx from the Saint John River (hereafter SJR) which ranges from about 500 m^3 s^{-1} during summer low-water conditions [29] to a maximum of about 10,000 m^3 s^{-1} during the spring freshet [30]. The combination of tidal flow and river runoff generates sharp density fronts which propagate through the vicinity of the port. Due to the large tidal amplitudes in the SJH and throughout the Bay of Fundy, there are significant inter-tidal zones (tidal flats) in the area. The local geography also leads to a complex flow system in the estuary. As SJR discharges into the harbour, water flows over a shallow sill into the Reversing Falls (see inset in Figure 1), a gorge in which rapids usually reverse in direction. Above the sill, the river expands into a river system that receives fresh water input from a large watershed that includes many lakes, hereafter called the river estuary. The Reversing Falls sill, between the harbour and the river estuary, constrains the upstream propagation of tides into the river estuary [28]. The flow conditions in the Reversing Falls change dramatically over the course of a tidal cycle, with the flow alternating between upstream and downstream. Typically, during the flood tide, the sea level in the harbour rises above the water level in the river estuary and there is a strong flow of saline water into the river system [31]. During the ebb tide, the sea level in the harbour drops to a lower level than in the river estuary and a mix of river and sea water flows back out into the harbour [31]. During freshet conditions, if river levels are sufficiently high, the river discharge dominates the flow through the Reversing Falls, such that there is no reversal of the flow. In the river estuary, a two layer system of fresher surface water and brackish bottom water, with a salinity of up to 20 ppt [28], is established in Long Reach (shown in Figure 1) and is present for most of the year, except during the spring and fall freshets when it may be flushed away [28,32]). In winter, the ice is present on the river estuary, but only thin ice appears near the shore in the harbour [28].

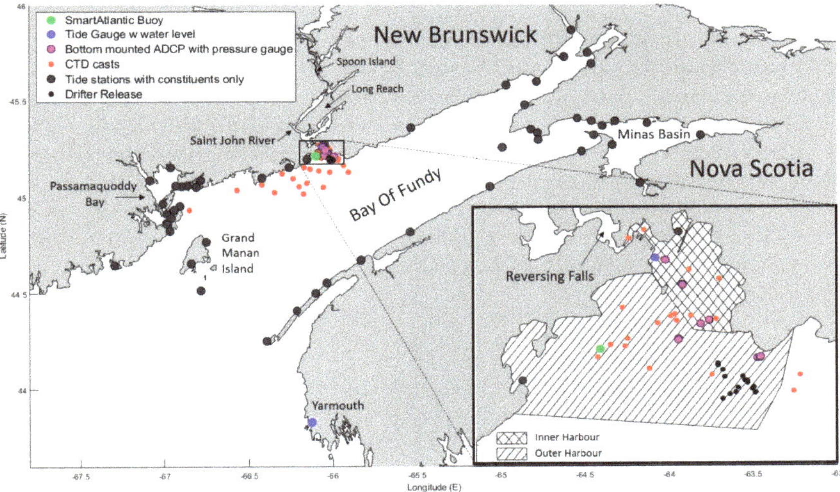

Figure 1. Map of the study area including the locations of the ocean observations, and the boundaries of the inner harbour and outer harbour.

3.2. Available Ocean Observations

A reason for choosing the Port of Saint John as the study area was the quantity and quality of available ocean observations during a one-year window, from 1 May 2015 to 30 April 2016, which was subsecuently set as the time period for the model simulation and evaluation. The types and locations of the available observations are shown in Figure 1. The map also divides the study area into three sub-areas: the inner harbour, the outer harbour, and the greater Bay of Fundy. Table 1 provides a summary of the observations, including the variables that were measured by each instrument and the number of observations in each of the three sub-areas. Note that extensive effort was put into collecting and carrying out the quality control on the observational data by various DFO projects.

Table 1. A summary of the observations that were used for the evaluation.

Instrument	Variable	Number of Observations		
		Inner Harbour	Outer Harbour	Greater BoF
SmartAtlantic Buoy	Sea Surface Temperature	0	1	0
Tide Gauge	Water Level	1	0	1
Historical Tide Gauges	Tidal Constituents	1	1	46
Moored ADCP	Currents	6	5	0
	Water Level			
CTD Profiles	Temperature	23	25	32
	Salinity			
Drifters	Surface Drifter Trajectories	0	134	0

All tide gauge and tide station data were obtained from the Canadian Hydrographic Service (CHS). The "constituent only" stations provided the harmonic constants for sets of tidal constituents based on the analysis of past water level time series observations, but did not provide observational time series during the evaluation period. The time series of water levels were available from two real-time tide gauges in the study area: one in the inner SJH and one at Yarmouth. The SJH gauge was critical for evaluating the tidal and non-tidal components of the modelled water level in the inner harbour. The Yarmouth gauge, located near the mouth of the Bay of Fundy, was used to assess the lateral open boundary condition and large-scale performance of the models.

DFO provided observations from 11 moored Acoustic Doppler Current Profilers (ADCPs) that were deployed between August 2015 and February 2016: six in the inner harbour and five in the outer harbour. The ADCPs were either moored a few metres off the bottom or within tripod bottom mounts and recorded the velocity in 1 m bins. To ensure confidence in the observations, a minimum of 75% ping return cut-off was used to define the valid observations. The ADCPs were also set to record the bottom pressure. The bottom pressure data were converted into variations of water level. The "inverse barometric" effect, accounting for the water level variations due to surface atmospheric pressure, was computed from the forcing data and added to the water level derived from the bottom pressure measurement.

Conductivity-temperature-depth (CTD) profiles were also provided by the DFO and were used to evaluate the modelled temperature and salinity fields. There were 80 CTD casts collected at 35 stations in and around SJH between June 2015 and April 2016. Although it is difficult to fully evaluate the large-scale models with point measurements, the locations and timings of the CTD casts made it possible to evaluate the seasonal changes in both the water properties and the depth of the mixed layer at representative locations in the three sub-areas. A SmartAtlantic [33] buoy located just outside the SJH provided the only time series measurement of the sea surface temperature. The buoy data, provided by the Fisheries and Marine Institute of the Memorial University of Newfoundland, were essential for evaluating the seasonal cycle of SST. Monthly mean SST and SSS (sea surface salinity) over the Bay of Fundy were evaluated in a qualitative manner to identify any large-scale discrepancies between the two models.

Lastly, the DFO deployed 134 surface drifters near SJH (in the outer harbour) between July 2015 and January 2016. Four different types of surface drifters were used: Seimac Accurate Surface Tracker barrel-shaped drifters (hereafter referred to as Barrel), MetOcean iSphere spherical drifters (hereafter referred to as Sphere), MetOcean CODE/Davis drifters (hereafter referred to as Davis), and Oceanetic Surface Circulation Tracker drifters (hereafter referred to as Sponge). The drifters provided records of drift trajectories over short periods of time (typically only one day). They were used to evaluate the drifter simulations based on the modelled surface current and the influence of surface winds (described in Section 3.7), and more generally, the models' ability to predict drift, which is one of the primary applications of the models under OPP.

3.3. Setup of NEMO and FVCOM Models

The models are based on NEMO version 3.6 and FVCOM version 3.2.1. Both models solve the 3D equations controlling the variations of ocean currents, water levels, temperature and salinity. The hydrostatic and Boussinesq approximations are applied. The minimum water temperature was set to freezing temperature. This is a simplified approach to account for sea-ice formation in winter, without turning on the sea-ice modules in either model.

The setup of the model domains considered the large-scale ocean forecasting system for providing the lateral open boundary condition (OBC). At the time as the evaluation, the available large-scale model was RIOPS which has a nominal horizontal resolution of 1/12-degree longitude/latitude, about 7.5 km in the study area. The finer coastal ocean prediction system for the east coast of Canada, at the nominal resolution of 1/36-degree, was still in the planning stages.

To take the OBC from RIOPS, FVCOM had the advantage of using a single unstructured-grid with variable horizontal resolution. Based on previous experience, the FVCOM domain was set as shown in Figure 2a, which encompasses the Bay of Fundy and Gulf of Maine, and extends offshore to include the Scotian Shelf and the shelf break. The horizontal cell size ranges from 14 km offshore to 48 m in SJH (Figure 2b). There are 21 geometrically spaced vertical sigma-levels resulting in layer thicknesses ranging from centimeters at the surface, to hundreds of meters at depth in the shelf break area.

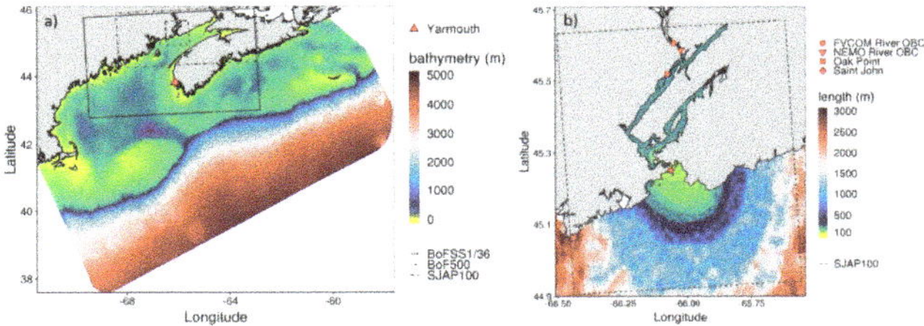

Figure 2. (**a**) Bathymetry over the FVCOM (Finite Volume Coastal Ocean Model) grid and the outlines of the Bay of Fundy and the western Scotian Shelf model component with a nominal horizontal resolution of 1/36-degree (BoFSS1/36) (solid line), the Bay of Fundy model component with an approximate 500 m resolution (BoF500) (dash-dot line), and the inner and outer Saint John Harbour model component with an approximate 100 m resolution (SJAP100) (dashed line) model domains. (**b**) Colormap showing the variable resolution of the FVCOM (as element side length in m) with the outline of the SJAP100 model domain (dashed line). Location of the river boundary, where the open boundary condition (OBC) is applied, is shown for the FVCOM (circle) and NEMO (Nucleus for European Modelling of the Ocean) (inverted triangle). Yarmouth (triangle), Saint John (diamond) and Oak Point (square) are marked for reference.

The structured-grid of NEMO does not have the flexibility to vary the horizontal grid size greatly with a single design. Instead, the approach taken was to develop three configurations (shown in Figure 2a) that are connected with one-way nesting. The grids of the three components are all aligned with the tri-polar ORCA grids created by the DRAKKAR Group [25,34]. The outer component covers the Bay of Fundy and the western Scotian Shelf with a nominal horizontal resolution of 1/36-degree, hence referred to as BoFSS1/36. The intermediate component covers the Bay of Fundy with an approximate 500 m resolution (hereafter BoF500). The inner component covers the inner and outer harbour with an approximate 100 m resolution (hereafter SJAP100). Note that the Saint John river estuary system, including the Reversing Falls, is included in SJAP100, roughly represented in BoF500, but totally excluded in BoFSS1/36. The one-way nesting is achieved by the larger domain component providing the OBC forcing to the next level smaller domain component successively, i.e., from RIOPS to BoFSS1/36, from BoFSS1/36 to BoF500, and from BoF500 to SJAP100. NEMO uses z-levels in the vertical space, with "bottom partial cells" for the accurate representation of the varying bathymetry, and the "variable volume level" scheme [35] to allow for the stretching and compression of the level thickness according to the changing water levels. Regarding the set-up of vertical levels, BoFSS1/36 uses the same 50-level setup as RIOPS but only has 28 active levels because its domain does not reach the deep ocean; BoF500 and SJAP100 have the same setup with 41 and 35 active levels, respectively, due to the difference in the maximum water depth between the two domains. The finest vertical resolution is near the surface, 1 m for all three components.

A high-resolution coastline dataset provided by the CHS was used to define the coastline represented in FVCOM and each component of NEMO. The model bathymetry within the Bay of Fundy and part of the Gulf of Maine was interpolated from a high-resolution dataset constructed from CHS and the Ocean Mapping Group, University of New Brunswick (OMG) data sources. Over the area not covered by the CHS/OMG data, NEMO used the global high-resolution SRTM30 bathymetry product [36] and FVCOM used bathymetry from the Scotia-Fundy-Maine grid of WebTide. As a wetting/drying scheme is unavailable in NEMO version 3.6, several modifications were made to the coastline and bathymetry, mainly in the upper Bay of Fundy and in the SJH which has large intertidal areas. Part of the intertidal zone was excluded, and part had the water depth increased to avoid drying and to maintain numerical stability. The modifications ensured the good representation of the M2 tides in and near the SJH [25].

Except for the river estuary system, which was initialized with the same data as the NEMO BoF500 simulation, the FVCOM model was initialized with the 3D temperature and salinity from the RIOPS solution on 1 February 2015, and was ramped up from rest (zero sea level and velocity) over a period of 36 h. The outer component of NEMO (BoFSS1/36) was initialized with the RIOPS solution on 1 February 2015, including 3D temperature and salinity, sea levels, and currents. BoF500 was initialized with the BoFSS1/36 solution on 1 April 2015, but with the temperature and salinity in the river estuary system taken from the solution of a separate 1-year simulation. Finally, SJAP100 is initialized on 10 April 2015, from the BoF500 solution. The simulations prior to 1 May 2015 (the start of the evaluation duration), are treated as the spin-up of the models. For FVCOM, the model spin-up occurs between 1 February and 30 April 2015. This spin-up period is considered sufficient for the winter conditions of the study area [25].

Input data for the lateral OBC included tidal and non-tidal components. For the FVCOM model and the BoFSS1/36 of NEMO, the non-tidal water levels for the OBC, at hourly intervals, were obtained from the de-tided RIOPS solution. The NEMO OBC also included non-tidal depth-averaged velocity from RIOPS. De-tiding is necessary because the RIOPS tidal solution contains significant errors in the Bay of Fundy. The input for OBC also included the daily averaged 3D temperature and salinity (without de-tiding) from RIOPS. The tidal water levels (and depth-averaged velocity for NEMO only) were computed from the harmonics of five major tidal constituents (M2, N2, S2, K1 and O1) from the WebTide solution [26]. Other tidal constituents will be included in the future development and operational application of the models. The BoFSS1/36 obtained a sufficiently accurate tidal solution near the open boundary of BoF500, and hence, BoF500 used both the tidal and non-tidal forcing for OBC from the solution of the BoFSS1/36, at 0.5 h intervals. Consequently, the SJAP100 took the full OBC forcing from the BoF500, also at 0.5 h intervals. Regarding the setup of the OBC, NEMO applied a "radiation" scheme for the barotropic (depth-averaged) current normal to the open boundary [37], and a "flow relaxation" scheme for temperature, salinity and the baroclinic current over a "relaxation zone" [38]. FVCOM enforced the water level, temperature and salinity at the open boundary nodes. A sponge layer was implemented at the open boundary with a radius of 25 km. Within the sponge layer, the longitudinal and latitudinal velocities, u and v, were reduced by a factor of u/(1+ Cspg u2) and v/(1+ Cspg v2). The friction coefficient, Cspg, was specified to be 0.002 at the open boundary nodes and linearly decreases to zero over the radius of the sponge layer.

The Saint John River runoff was included in FVCOM, and the BoF500 and SJAP100 components of NEMO. For both the NEMO and FVCOM configurations, the open boundary of the river was placed upstream of the Long Reach and below the Spoon Island (see Figure 2). This location is close to the upstream extent of the salt wedge intrusion and limits the inclusion of complicated river morphology.

The river forcing was not included in the BoFSS1/36, but this did not cause a significant error in the BoFSS1/36 solution at the lateral open boundary of the BoF500. Although both models have the ability to include river discharge, suitable input data were not available, and thus the entry of the SJR was treated as an open boundary forced with the observed water levels at the Oak Point station in the river estuary system (Figure 2). A constant value of 0.7 m was subtracted from the observed Oak Point water levels prior to using them for OBC forcing. This constant is the estimated difference in the reference levels between the Oak Point station and the model [21].

At the sea surface, the models were forced using the forecast fields from the High-Resolution Deterministic Prediction System of the CCMEP, with a horizontal resolution of 2.5 km [39]. The forecast variables included hourly winds at 10 m height, air temperature and specific humidity at 2 m height, sea-level air pressure, precipitation, and surface incoming longwave and shortwave radiation. In NEMO, the surface wind stress, the sensible and latent heat fluxes, and the rate of evaporation were calculated using the forecast variables and the bulk formulae of the Coordinated Ice-Ocean Reference Experiments [40]. In FVCOM, the corresponding calculations were based on the Coupled Ocean-Atmosphere Response Experiment (COARE) version 3.0 algorithm [41], which was modified to use the specific humidity (instead of the relative humidity) as the input. Note that for the solutions provided for evaluation, the forcing of surface freshwater flux, due to evaporation and precipitation, was included in NEMO but not in FVCOM. According to the evaluation results (presented in the following sections), this difference was not a main factor for the difference in the model solutions.

Finally, both NEMO and FVCOM include parametrizations for unresolved sub-grid processes. The parameterizations used by the two models are similar but with some subtle differences, and are tuned separately to improve the performance of the models. NEMO uses a partial slip scheme for velocity near the lateral solid boundary, and the variable horizontal mixing for momentum computed with the scheme of Smagorinsky [42,43]. FVCOM specifies a zero normal component of the velocity on the lateral solid boundary and also uses the Smagorinsky eddy parameterization for the horizontal diffusion [44]. Both models adopt high-order closure schemes to compute the vertical mixing for tracers and momentum [45]: the k-ε configuration of the generic length scale scheme for NEMO and the 2.5 level scheme of Mellor-Yamada for FVCOM. Both models adopt the quadratic law for bottom drag parameterization but with different tuning of the drag coefficient to maintain the numerical stability and improve the tidal solutions. For NEMO, the drag coefficient has background values of 2.5×10^{-3}, 4×10^{-3} and 5×10^{-3} for the BoFSS1/36, BoF500 and SJAP100 components, respectively. In BoF500 and SJAP100, the drag coefficient was locally increased in the upper Bay of Fundy and in the Reversing Falls [46]. In FVCOM, the coefficient Cd is calculated by fitting a logarithmic bottom boundary layer to the model at a specified height zab above the bottom with a bottom roughness scale of z0, Cd = max(k2/ln(zab/z0)2,Cdmin) where k is the von Karman constant, z0 = 0.001, and Cdmin = 0.02. To obtain stable runs and simulate the strong dissipative effects of the rapids in the Reversing Falls, the sponge layer feature of FVCOM was used to reduce the horizontal currents in the Falls and approximately 1 km downstream using a damping coefficient of 0.015 m and a sponge radius of 500 m.

3.4. Evaluation of the Tidal Water Level and Currents

The tides dominate the variations of water levels and currents in the BoF and are a major concern for navigational safety in SJH. The tidal currents have a direct impact on the drift of spilled and other hazardous materials in the water.

The evaluation of the tidal water level and currents was based on their harmonic constants using the following metrics:

Difference in amplitude and phase:

$$D = X_m - X_o \tag{1}$$

Percent difference in amplitude:

$$D_\% = \frac{X_m - X_o}{X_o} \times 100\% \tag{2}$$

Vector difference for water level:

$$D_V = \left| A_o e^{i\varphi_o} - A_m e^{i\varphi_m} \right| \tag{3}$$

Vector difference for currents:

$$D_V = \frac{1}{T} \int \left| Q_o(t) - Q_m(t) \right| dt \tag{4a}$$

with:

$$Q(t) = u + iv = M[\cos(\omega t - \psi) + \varepsilon i \sin(\omega t - \psi)] e^{i\theta}. \tag{4b}$$

In the above equations, the subscript "o" denotes the observations and "m" denotes the model results. All the variables are defined for a single constituent. In Equations (1) and (2), X represents either the amplitude or phase. In Equation (3), A and φ denote the amplitude and phase of water level, respectively [26]. In Equations (4a) and (4b), Q is the 2D tidal flow (u, v) represented in complex form (with $i = \sqrt{-1}$), t is the time and T is the period of a specific tidal constituent with frequency ω. The other symbols are related to the tidal current ellipse: M is the length of the semi-major axis, ε is the eccentricity (defined as the ratio of the minor axis to the major axis), θ is the orientation, and ψ is the phase of the complex current [46].

For an area like the BoF, where there is significant spatial variability in the tides, percent difference is a more useful metric than the absolute difference. Vector difference (Equations (3) and (4)) combines the errors in amplitude and phase into one metric. Vector difference was the primary metric for evaluating the models' overall skill in reproducing the tides, but evaluating the amplitude and phase separately was useful for understanding the root of the errors.

Harmonic analysis on the water level and currents used the t_tide package for [47] with the Rayleigh criteria equal to 1. For stations with water level observations during 1 May 2015–30 April 2016, which include the real-time gauges and the moored ADCPs, the model output was extracted for the same time period as the observations. We also obtained harmonic constants derived from historical observations at the "constituent-only" stations. At these stations, the harmonic constants of the model solutions were obtained by performing the analysis on the full 1 year model output, regardless of the length of observations that were used to compute the observed constituents. For tidal currents, the model output was interpolated to the observed depths. Constituents (major and minor axis, inclination and phase) were computed for each depth bin of the observations. The evaluation focused on selected depths: the "surface", and 5, 10 and 20 m from the surface. Here, the "surface" is the uppermost bin with >75% ping return for the entire time series.

The polar plots in Figure 3 visually report the tidal water level metrics for each station in the inner harbour (Figure 3a) and outer harbour (Figure 3b), for the M2 constituent. The tidal ellipses in Figure 3 are representative examples of the M2 current at the surface from one station in the inner harbour (Figure 3c) and one station in the outer harbour (Figure 3d). Table 2 summarizes the evaluation results for the M2 constituent, for several selected metrics: D% (amplitude), D (phase) and Dv for water levels; and D (major axis), D (phase), and Dv for currents. The mean and the standard deviation of the mean for each metric, across the stations in the inner harbour and outer harbour, separately, are listed. Note that the means and standard deviations were computed using the absolute value of the metric so that the values reflect the average error in the model, regardless of whether the error was positive or negative. Similar tables were produced for other tidal constituents, but, for brevity, they are not presented here.

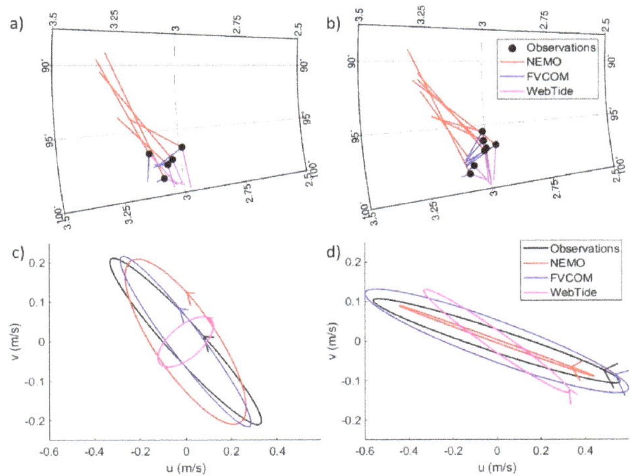

Figure 3. (**a**,**b**) Polar plot of the amplitude (m) and phase (degrees) of the observations (black circles) versus the model results in the inner harbour (**a**) and outer harbour (**b**). The size and direction of the vectors (NEMO, red; FVCOM, blue; and WebTide, purple) indicate the difference in the amplitude and phase between the observations and the model. Regional ice-ocean prediction system (RIOPS) results were omitted from these plots. (**c**,**d**) Observed and modelled M2 tidal ellipses for the surface currents from one station in the inner harbour (**c**) and one in the outer harbour (**d**). Position of the arrow on the ellipse indicates the phase, and the direction of the arrows indicates the counter-clockwise rotation of the complex current.

Table 2. The evaluation metrics for the M2 tidal water level and currents (see definitions in Equations (1)–(4) and in the text), averaged for the stations in the inner and outer Saint John Harbour (SJH). Numbers in brackets are the standard deviations across the stations. The metrics are computed for the NEMO, FVCOM, RIOPS and the WebTide solutions, separately. Note that for D and $D_\%$, the mean and standard deviations are calculated using the absolute values of the D and $D_\%$ values at each station.

		NEMO	FVCOM	RIOPS	WebTide
	Water Level (M2)				
Inner Harbour	$D_\%$ (amplitude)	7.68 (1.59)	2.48 (1.20)	21.78 (2.38)	0.93 (0.94)
	D (phase, degrees)	5.15 (2.69)	1.03 (0.67)	56.31 (0.80)	2.48 (1.09)
	D_V (m)	0.38 (0.12)	0.09 (0.04)	2.61 (0.01)	0.14 (0.04)
Outer Harbour	$D_\%$ (amplitude)	6.99 (1.17)	1.89 (1.08)	21.85 (2.55)	1.43 (1.21)
	D (phase, degrees)	6.03 (2.80)	0.72 (0.88)	55.58 (0.82)	2.07 (0.97)
	D_V (m)	0.40 (0.14)	0.08 (0.04)	2.60 (0.07)	0.12 (0.04)
	Tidal Current (M2)				
Inner Harbour	D (maj. axis, m s^{-1})	0.04 (0.03)	0.03 (0.02)	-	0.15 (0.16)
	D (phase, degrees)	38.77 (36.40)	13.21 (6.19)	-	72.05 (46.56)
	D_V (m s^{-1})	0.08 (0.03)	0.07 (0.03)	-	0.18 (0.09)
Outer Harbour	D (maj. axis, m s^{-1})	0.16 (0.10)	0.04 (0.03)	-	0.25 (0.09)
	D (phase, degrees)	7.65 (4.04)	7.95 (4.77)	-	23.40 (8.11)
	D_V (m s^{-1})	0.13 (0.06)	0.08 (0.03)	-	0.21 (0.06)

For the tidal water levels, according to Dv, FVCOM performed the best overall, followed by WebTide, NEMO, and then RIOPS. The performance of the models was not significantly different between the inner harbour and the outer harbour. Compared to WebTide, FVCOM had a slightly larger error in amplitude but a smaller error in phase. The larger error in NEMO was due to an error in the input files that was identified toward the end of the evaluation and corrected afterward [25]. RIOPS significantly underperformed compared to the other models, mainly due to the coarse spatial resolution that cannot capture the resonance characteristics of the M2 tide in the Bay of Fundy and uncalibrated tides in general.

For tidal currents, according to the Dv metric, FVCOM and NEMO performed similarly in the inner harbour, but FVCOM performed slightly better than NEMO in the outer harbour. FVCOM had smaller errors in the phase in the inner harbour, and smaller errors in the major axis in the outer harbour. Both FVCOM and NEMO significantly outperformed WebTide, mainly because the WebTide does not include the baroclinic dynamics. The errors for the RIOPS currents were not quantified but were expected to be significantly larger.

3.5. Evaluation of the Non-Tidal (Residual) Water Level and Currents

The non-tidal (or residual) component of the water level is important for predicting extreme water level events like storm surges. Residual currents dominate the net (after averaging out tidal excursions) evolution of the drift trajectory and contribute to extreme events of concern for navigational safety.

The residual water level and currents were computed by removing the tidal component (as determined by t_tide) from the observations and model output. An additional filter was applied to remove the energy in the tidal period bands (22–28 h, 11–14 h, 5–7 h). For the currents, all energy at periods < 7 h was removed.

The following metrics were used to evaluate the residual water level and the currents at each station (and depth for currents):

Mean bias:

$$\overline{D} = \overline{X_m} - \overline{X_o} \tag{5}$$

Root mean square error:

$$RMSE = \left\{ \frac{1}{N} \sum_{i=1}^{N} (X_m - X_o)^2 \right\}^{1/2} \tag{6}$$

Relative average error:

$$RAE = 100\% \frac{\sum_{i=1}^{N} (X_m - X_o)^2}{\sum_{i=1}^{N} \left(\left| X_m - \overline{X_o} \right|^2 + \left| X_o - \overline{X_o} \right|^2 \right)} \tag{7}$$

Correlation:

$$R = \frac{\sum_{i=1}^{N} \left(X_m - \overline{X_m} \right) \left(X_o - \overline{X_o} \right)}{\left[\sum_{i=1}^{N} \left(X_m - \overline{X_m} \right)^2 \left(X_o - \overline{X_o} \right)^2 \right]^{\frac{1}{2}}} \tag{8}$$

Skill:

$$Skill = 1 - \frac{\sum_{i=1}^{N} \left| X_m - X_o \right|^2}{\sum_{i=1}^{N} \left(\left| X_m - \overline{X_o} \right| + \left| X_o - \overline{X_o} \right| \right)^2} \tag{9}$$

In the above equations, X denotes the time series of the water level or a velocity component, and the overbar denotes its time mean. The other notations are consistent with those used in Equations (1)–(4). In conjunction with the above metrics, we also compared the power spectra of the time series.

Figure 4 presents the observed and modelled time series of the residual water level at the Saint John tide gauge location and the associated spectra. The shaded areas in Figure 4a–c highlight the time frame of the Fall freshet (September 2015) and the Spring freshet (April 2016), which are shown in detail in Figure 4d,e, respectively. NEMO and FVCOM obtained consistent results for the two freshet events, including the deficiencies after the peak water levels. That is, the modelled water levels were lower than those observed after the peak of the fall freshet, and higher than that observed after the peak of the spring freshet. The causes of these discrepancies were not further explored. The three models (NEMO, FVCOM and RIOPS) all captured the main characteristics of the observed variations of residual water level over a full year period, including similar spectral distributions.

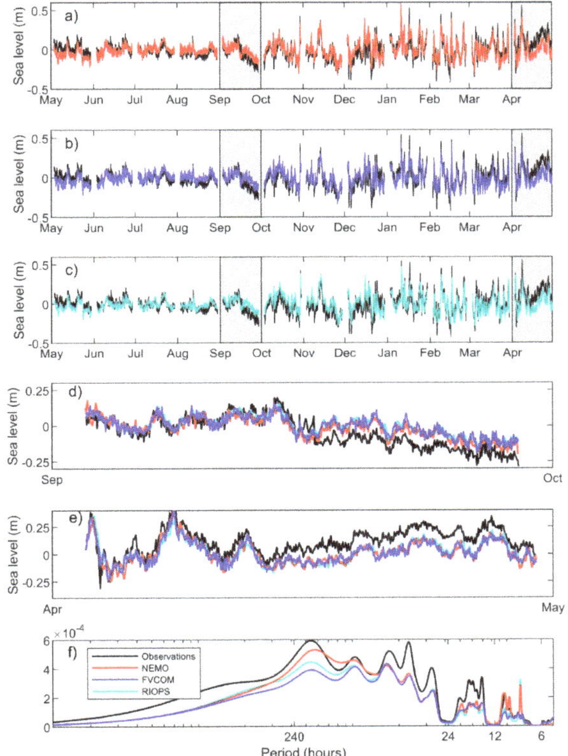

Figure 4. (**a–c**) One year time series of the residual water level at the Saint John tide gauge location from (**a**) the NEMO, (**b**) the FVCOM, and (**c**) the RIOPS. Observations are shown in black in each panel. The shaded areas in (**a–c**) highlight the time frame of the Fall freshet (September 2015) and the Spring freshet (April 2016), and are expanded in (**d**) and (**e**), respectively. (**f**) The spectra for the time series in (**a–c**) in variance-conserving form. The legend in panel (**f**) applies to all panels. The residual water level was computed by removing the tidal component (as determined by t_tide). An additional filter was applied to remove the extra energy in the tidal period bands (22–28 h, 11–14 h, 5–7 h).

Table 3 summarizes the results as the mean and standard deviation across all stations (and the depths for currents) for each metric. For the residual water level, the metrics were computed on the demeaned times series. The mean bias for the water level across all the stations (including the ADCP measurements) was not computed. NEMO, FVCOM and RIOPS showed very similar values for the three metrics (RMSE, RAE and correlation) listed. There are also no significant differences in the skills between the inner and outer harbours. The correlation of both models with the observed water level was high; on average, 0.74 in the inner harbour, and 0.78 in the outer harbour, but as much as 0.9 at some stations. The metrics showed no significant difference in model performance between the inner harbour and the outer harbour, but the spectra from the Saint John tide gauge indicates that, in the inner harbour, the NEMO model had better representation of the energy in the 5–10 day period band, and the FVCOM captured more of the high-frequency (<6 h) energy.

Table 3. The evaluation metrics for the residual water level and currents (see definitions in Equations (5)–(8) and in text), averaged for the stations in the inner and outer SJH, separately. Numbers in brackets are the standard deviations across the stations. RMSE, RAE, R and Skill are defined in Equations (6)–(9). Note that for the water levels, the mean bias is not computed and the other metrics are computed for the demeaned time series. Water level metrics are computed for NEMO, FVCOM and RIOPS. Current metrics are computed for NEMO and FVCOM only.

			NEMO	FVCOM	RIOPS
		Residual Water Level			
Inner Harbour		RMSE (m)	0.08 (0.02)	0.08 (0.02)	0.08 (0.02)
		RAE (%)	26.60 (14.85)	27.78 (13.33)	29.41 (12.32)
		R	0.74 (0.15)	0.74 (0.13)	0.73 (0.12)
		Skill	0.85 (0.09)	0.84 (0.08)	0.82 (0.08)
Outer Harbour		RMSE (m)	0.09 (0.03)	0.09 (0.03)	0.10 (0.03)
		RAE (%)	26.66 (17.83)	28.57 (18.28)	28.17 (17.13)
		R	0.78 (0.12)	0.78 (0.12)	0.77 (0.12)
		Skill	0.85 (0.11)	0.83 (0.12)	0.84 (0.11)
		Residual Currents			
Inner Harbour	u	Mean Bias (m s^{-1})	−0.05 (−0.05)	−0.04 (0.04)	-
		RMSE (m s^{-1})	0.08 (0.02)	0.07 (0.03)	-
		RAE (%)	77.68 (13.46)	73.33 (17.19)	-
		R	0.42 (0.19)	0.47 (0.23)	-
		Skill	0.51 (0.10)	0.51 (0.10)	
	v	Mean Bias (m s^{-1})	0.04 (0.07)	0.02 (0.05)	-
		RMSE (m s^{-1})	0.08 (0.04)	0.07 (0.02)	-
		RAE (%)	68.75 (14.28)	74.12 (10.71)	-
		R	0.53 (0.17)	0.49 (0.20)	-
		Skill	0.58 (0.10)	0.55 (0.05)	
Outer Harbour	u	Mean Bias (m s^{-1})	−0.04 (0.02)	−0.08 (0.02)	-
		RMSE (m s^{-1})	0.07 (0.01)	0.10 (0.02)	-
		RAE %	80.82 (11.85)	88.69 (10.08)	-
		R	0.26 (0.12)	0.32 (0.14)	-
		Skill	0.52 (0.08)	0.50 (0.06)	
	v	Mean Bias (m s^{-1})	0.06 (0.03)	0.00 (0.02)	-
		RMSE (m s^{-1})	0.07 (0.02)	0.04 (0.01)	-
		RAE (%)	95.08 (7.22)	82.45 (9.33)	-
		R	0.15 (0.10)	0.23 (0.11)	-
		Skill	0.38 (0.06)	0.49 (0.06)	

Figure 5 presents examples of the observed and modelled residual current at the surface from representative ADCPs in the inner harbour (a,b) and the outer harbour (c,d). As reflected in the figures, both NEMO and FVCOM captured the timing of large-scale events but underestimated the magnitude of the currents in the inner harbour. The magnitude of the currents was better predicted by the models in the outer harbour. The metrics in Table 3 indicate that both NEMO and FVCOM showed a better performance in the inner harbour (with average correlations in u an v between 0.42 and 0.53, respectively) compared to the outer harbour (average correlation between 0.15 and 0.32). We did not evaluate the models against RIOPS because its coarse horizontal resolution poorly represents the spatial variations of currents in SJH.

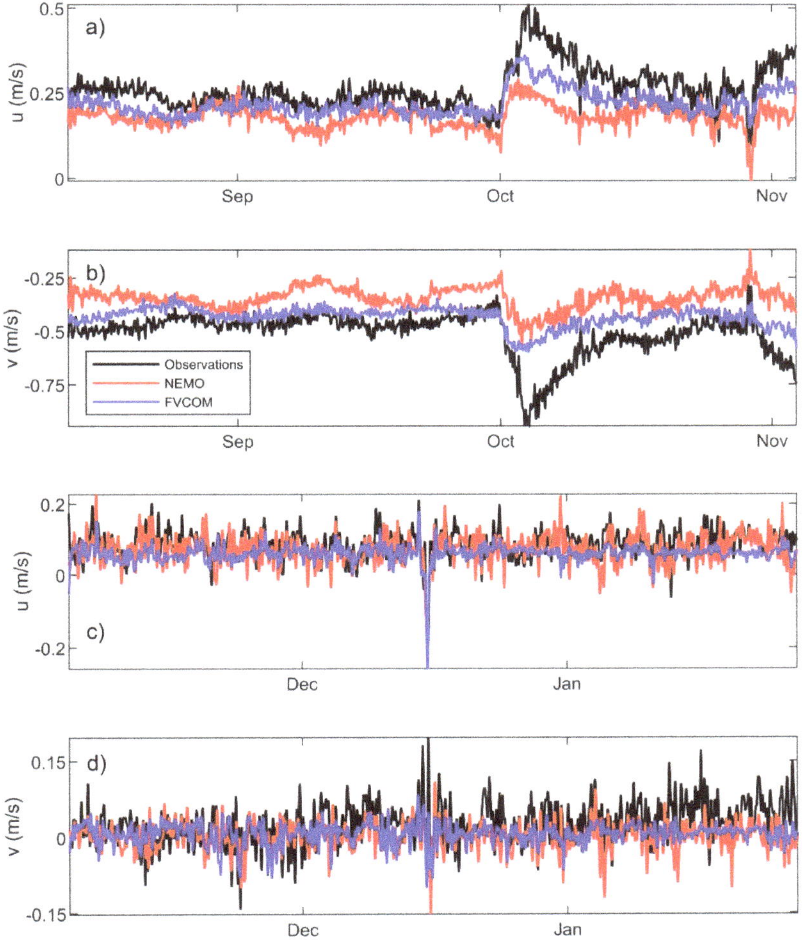

Figure 5. Observed (black) and modelled (NEMO, red; FVCOM, blue) time series of the u and v components of the residual currents, at the representative stations in the inner harbour (**a,b**) and outer harbour (**c,d**). The residual currents were computed by removing the tidal component (as determined by t_tide). An additional filter was applied to remove the extra energy in the tidal period bands (22–28 h, 11–14 h) and all energy at periods < 7 h.

3.6. Evaluation of the Temperature and Salinity Fields

Variations in the water temperature and salinity cause changes in density, density-driven currents, the stratification of the water column, and the chemistry (which subsequently influences the fate and behavior of spilled oil) of the water.

Variations in the temperature and salinity in SJH were first evaluated qualitatively for three cases: low water, spring freshet and fall freshet. Salinity variations are mainly attributed to the input of freshwater from the Saint John River and the mixing of freshwater and seawater in the harbour. Because NEMO and FVCOM were forced by the observed water level at Oak Point station (in the river estuary system), we first qualitatively diagnosed the freshwater flux (river runoff) across a section a few kilometers downstream of Oak Point and determined that the diagnosed fluxes were similar in both models. Figure 6 presents the surface density fields from NEMO and FVCOM, as well as satellite-derived images of suspended particulate matter (provided by E. Devred, DFO), during two different phases of the tide. These figures show that the models obtained a similar extent of the river plume, consistent with the interpretation of satellite images.

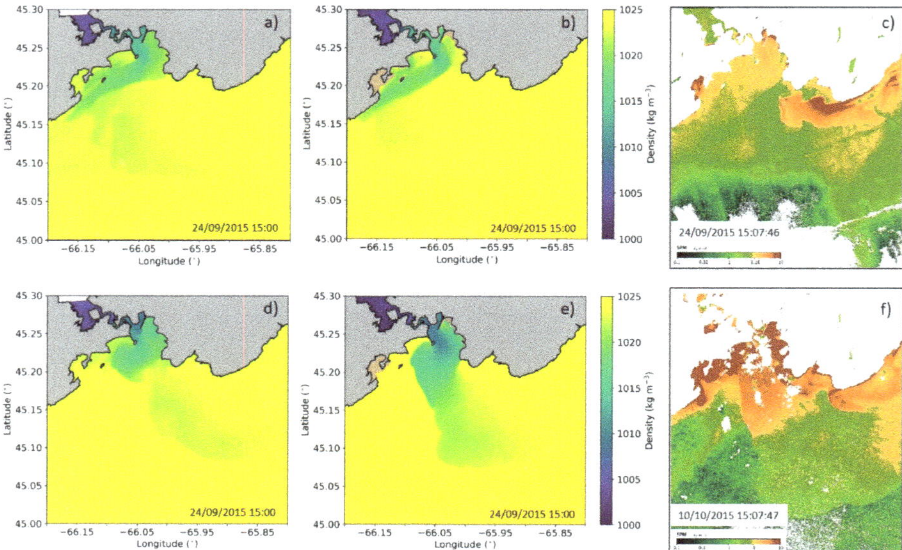

Figure 6. Snapshot of the surface density (kg m^{-3}) from the NEMO (**a,d**), and the FVCOM (**b,e**) within 15 min of the satellite-derived images of suspended particulate matter (**c,f**) provided by E. Devred, Fisheries and Oceans Canada (DFO). The dates and times of the satellite images, as well as the modelled surface density field are indicated.

The SmartAtlantic Buoy provided the only time series measurement of SST in SJH and was essential for evaluating the annual cycle of the SST. Figure 7 presents the comparison of the observed and modelled SST from the NEMO and FVCOM. The mean bias, RMSE, RAE, correlation and skill (as described in Equations (5)–(9)) were computed for the time series of SST from NEMO, FVCOM and RIOPS, and are presented in Table 4. Because the temperature fields from RIOPS were only available as daily means, the metrics for RIOPS were computed using the daily mean of the observations. Both NEMO and FVCOM captured the observed seasonal cycle as well as the high-frequency variations, but there was a warm bias in the FVCOM model. FVCOM performed slightly better than NEMO in terms of correlation (0.96 for FVCOM and 0.94 for NEMO), and both models outperformed RIOPS across all metrics except mean bias.

Figure 8 compares the observed vertical profiles of temperature and salinity from the CTD casts with the model simulations at two representative stations: one in the inner harbour (a,b), and one in the outer harbour (c,d). The temperature and salinity profiles fluctuate significantly with the phase of the tide in SJH, particularly for salinity due to the interaction of river runoff, tidal flow, and mixing. The observed profiles were compared to the nearest model output in time (t) and space. The figures also include shaded areas that represent the profiles within t − 3 h and t + 3 h which show how the modelled profiles varied over a 6 h period. As expected, the figures show more variation in the modelled T and S profiles in the inner harbour than in the outer harbour. For the qualitative evaluation of the profiles, the model results were "acceptable" if the observed profile fell within the shaded area. Generally speaking, NEMO showed a better representation of the vertical stratification. FVCOM obtained more uniform temperature and salinity profiles, probably due to too strong vertical mixing that needs to be tuned.

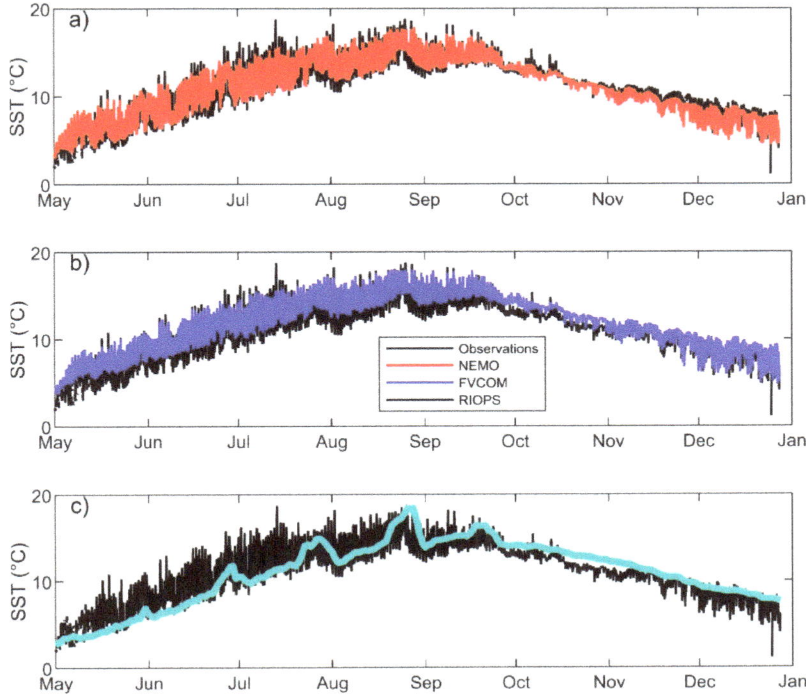

Figure 7. Time series of the sea surface temperature (SST) from (**a**) the NEMO, (**b**) the FVCOM, and (**c**) the RIOPS compared to the observed SST at the SmartAtlantic Buoy. RIOPS times series is based on daily output.

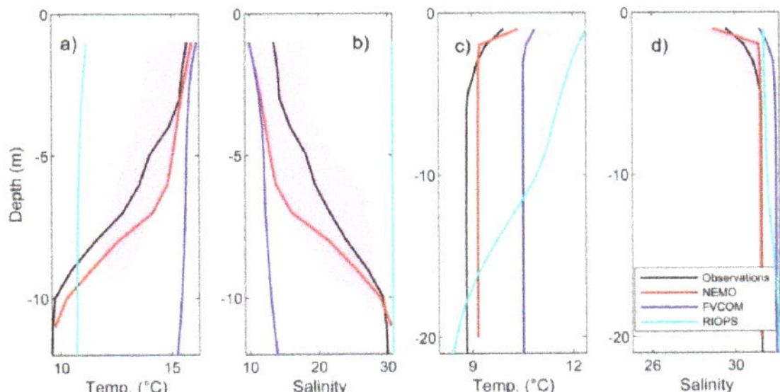

Figure 8. Observed (black) and modelled (NEMO, red; FVCOM, blue; RIOPS, cyan) temperature (**a,c**) and salinity (**b,d**) profiles from the representative stations in the inner harbour (**a,b**) and outer harbour (**c,d**). The model profiles are taken at the nearest hour to the observations, h. The shaded area around the NEMO and FVCOM profiles (light pink for the NEMO and light blue for the FVCOM) represents the profiles within $t - 3$ h and $t + 3$ h to show how the modelled profiles vary over a 6 h period. Where the areas overlap, the shading appears more purple.

The mean bias, RMSE, and RAE were computed for each profile, and the correlation and skill were computed using all the profile data in the inner harbour and outer harbour, separately. Table 3 summarizes the results, listing the mean and standard deviation across all stations in the inner harbour and outer harbour, separately, for mean bias, RMSE, and RAE. According to the correlation metric, NEMO and FVCOM performed better in the inner harbour than the outer harbour. With respect to salinity, the summarized metrics indicate that the errors for NEMO were lower than FVCOM, but a t-test shows no significant difference. The correlation metric indicates that NEMO had a slight advantage over FVCOM in the inner harbour, while the opposite was true in the outer harbour.

Table 4. Evaluation metrics for the time series of the SST (compared to the observations from the SmartAtlantic (SA) Buoy), and the water temperature and salinity (compared to the observed CTD profiles). See the definitions of the metrics in Equations (5)–(9) and in the text. The mean bias (D), RMSE and RAE are averaged for the stations in the inner and outer SJH. Numbers in brackets are the standard deviations across the stations. Correlation and skill were computed using all the profile data in the inner and outer harbour, separately.

		NEMO	FVCOM	RIOPS
		SST (time series)		
SA Buoy	D (°C)	0.08	0.84	0.00
	RMSE (°C)	1.05	1.23	1.63
	RAE (%)	5.62	7.82	11.81
	R	0.94	0.96	0.89
	Skill	0.97	0.96	0.94
		Water temperature		
Inner Harbour	D (°C)	0.22 (0.55)	1.62 (0.72)	0.06 (1.91)
	RMSE (°C)	0.70 (0.32)	1.76 (0.90)	1.64 (1.36)
	RAE (%)	43.03 (38.76)	82.12 (20.20)	90.23 (9.08)
	R	0.98	0.96	0.80
	Skill	0.98	0.91	0.89
Outer Harbour	D (°C)	0.21 (0.45)	1.47 (0.34)	0.86 (0.59)
	RMSE (°C)	0.50 (0.44)	1.56 (0.27)	1.13 (0.48)
	RAE (%)	55.28 (30.08)	90.75 (16.47)	93.35 (7.17)
	R	0.99	0.99	0.98
	Skill	0.99	0.95	0.98
		Salinity		
Inner Harbour	D (psu)	−0.89 (1.70)	−2.04 (3.04)	5.80 (7.33)
	RMSE (psu)	2.28 (1.46)	3.28 (3.15)	7.48 (7.01)
	RAE (%)	29.00 (31.49)	44.13 (36.87)	98.80 (1.12)
	R	0.97	0.93	0.08
	Skill	0.98	0.96	0.44
Outer Harbour	D (psu)	−0.06 (0.96)	0.76 (0.94)	0.72 (1.17)
	RMSE (psu)	1.31 (1.32)	1.45 (1.30)	1.92 (1.82)
	RAE (%)	39.54 (30.47)	50.65 (31.13)	97.35 (3.45)
	R	0.75	0.79	0.33
	Skill	0.86	0.86	0.32

The high-variability of temperature and salinity in Saint John Harbour, particularly when close to the mouth of the river, was difficult to evaluate with point measurements, but nonetheless, the analysis revealed the presence of a warm/salty bias in the FVCOM model. The warm/salty bias in FVCOM has since been corrected and was due in small part to a technical issue in the implementation of the COARE 3.0 algorithm that was identified toward the end of the evaluation, and in large part due to the placement of the outer boundary extending past the shelf break. That being said, both models performed better than RIOPS due to the inclusion of the freshwater flux from the Saint John River.

3.7. Evaluation of Drifter Simulations

Improving drift prediction is one of the primary objectives of the oceanography component of the OPP. Hence, our evaluation process includes a simulation of the surface drift trajectories and a comparison with the observed trajectories of the four types of surface drifters deployed in the SJH.

Drifter simulations were performed using the Canadian Oil Spill Modelling Suite (COSMoS) [48]. COSMoS is a Lagrangian displacement model that solves the equations of motion for drift using surface currents and surface wind. It takes surface currents from the results of models on the structured or unstructured native grids. Note that currents are computed at the centre of the FVCOM grid elements so that the solutions had to be interpolated to the nodes. The computation of the drifter trajectories uses a fourth order Runge-Kutta scheme. The simulations were run with surface currents from the NEMO (SJAP100 and BoF500), FVCOM, and WebTide. A simulated drifter was "released" at the same time and location of each observed drifter, and the simulation was terminated at the end of the observed drifter release. The drift evaluation was restricted to the smallest common domain, which is the SJAP100 domain. Even though BoF500 has a lower resolution in this area, the drift simulations were completed using currents from both SJAP100 and BoF500 to help understand the influence of increased resolution on drift prediction accuracy.

The inclusion of surface wind in the drifter simulations is to account for supplemental wind effects on the drifting object that are not included in the surface currents. However, the true wind effects may differ for different types of drifters. Simulations were run with 0% and 3% of the wind speed (10 m winds from HRDPS) to evaluate the effect. The results consistently showed that adding 3% of wind to the drifter simulations slightly improved the overall skill of the NEMO-based trajectories, but degraded the FVCOM-based solutions. Thus, the results presented here focus on the solutions with no added wind effect.

The first aspect of the drift evaluation was to qualitatively compare the distributions of surface velocities derived from the observed and modelled drifter trajectories. Drifter velocity is approximated as a forward Euler difference between the successive positions and reported in along-shore (241 °T positive) and cross-shore (151°T positive) components. Note that without including the wind effect, the "drifter velocity" from the models are very close to the surface currents. Figure 9a–e shows the distributions of drifter velocities for the Davis drifters, but the results were similar for all drifter types. The mean and RMS speeds derived from these distributions were within a few percent of each other for all cases, and the shapes of the distributions were generally similar. However, the distribution of the observed drift velocities appears to have heavier tails than the modelled drift velocity distributions, which is consistent with the underprediction of currents noted in Section 3.5. As expected, the distributions based on WebTide were more bi-modal (consistent with ebb and flood tide) in the along-shore direction and more peaked around 0 in the cross-shore direction. They showed no mean speed, and underpredicted the variability in the drift speeds. Such consequences of WebTide having no baroclinic flow was evident here, as well as across all the evaluations of the drifter trajectories (panels d, i, n, s, w, aa of Figure 9). Therefore, the WebTide-based results are not discussed further in this paper.

The second aspect of the drift evaluation was to compare the absolute dispersion of the observed and modelled drifters by plotting distributions of along-shore and cross-shore drifter displacements as a function of time lag, τ. Mathematically:

$$\Delta X(\tau) = X(t - \tau) - X(t) \tag{10}$$

where ΔX is the change in along-shore or cross-shore displacement during time lag τ. To account for all the possible displacements in the area sampled by the drifters, all non-overlapping segments of drifter tracks were used in the analysis. This is equivalent to assuming homogeneous and stationary absolute dispersion statistics.

Figure 9f–o shows the dispersion of the observed and modelled Davis drifters for up to 24 h. Displacements in the along- and cross-shore directions (y axis) are binned at hourly intervals (x axis) and the density of the observations in each bin is plotted to visualize the temporal evolution of the drifter 'cloud'. The statistics of along-shore dispersion display a clear 'tidal' character, that is, the expansion and contraction of the 'displacement cloud' on a ~12 h cycle. This is superimposed on a mean down-coast (along-shore positive) drift. Oscillations were not as evident in the cross-shore displacements. These characteristics were consistent across all four drifter types.

Generally, the distributions of displacement produced by the NEMO and the FVCOM were similar to the observed statistics, though extreme displacements vary slightly between models and observations. Both FVCOM and NEMO retained the general shape of the tracks, but on average the modelled along-shore displacements were lower than the observed values, indicating that the motion was characteristically similar but occurred closer to Saint John in the models than was observed. In the cross-shore direction, both the FVCOM and the NEMO predicted a larger mean displacement than was observed, indicating that the modelled drifters travelled further offshore than the observed ones.

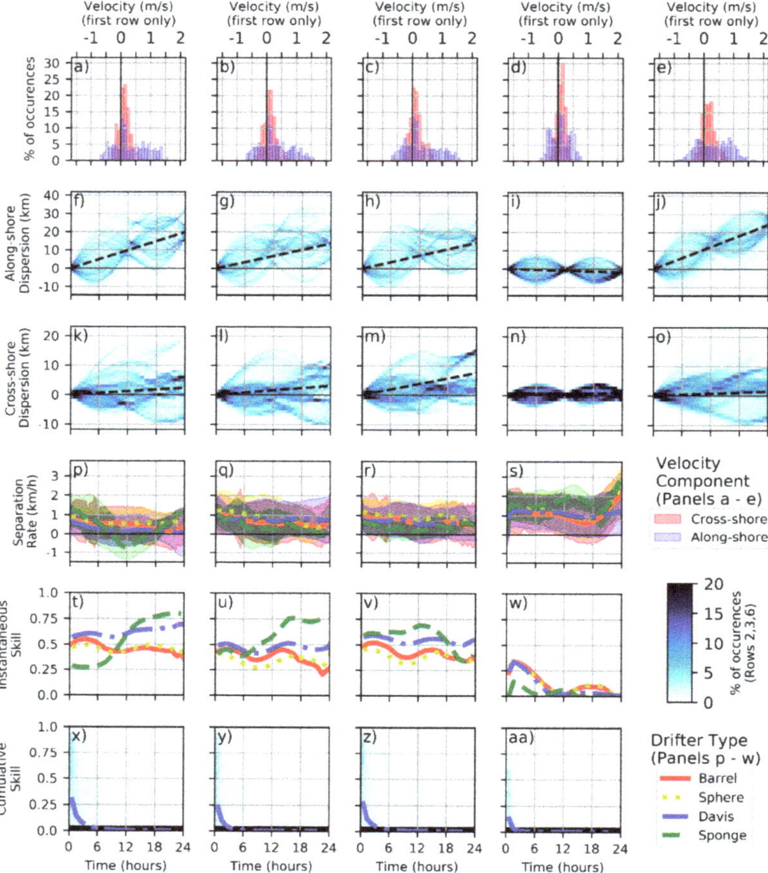

Figure 9. Results of the drifter analysis for the simulated drifters using the FVCOM (left column), SJAP100 (second from left), BoF500 (middle), WebTide (second from right), and observations where applicable (right column). Distributions of along-shore (blue) and cross-shore (red) drifter velocity (for the Davis drifters) are shown in the first row (**a–e**). Absolute dispersion in the along-/cross-shore directions (for the Davis drifters) is shown in the second and third row (**f–o**) with the mean drift superimposed on the density of the simulated/observed positions in each direction. Separation rate between the simulated drifters and the corresponding observations is shown in the fourth row (**p–s**) for all four drifter types. Instantaneous skill corresponding to the various model predictions is shown in the fifth row (**t–w**) for all drifter types, and the cumulative skill for the prediction of the Davis drifter trajectories is shown in the sixth row (**x–aa**), with the mean cumulative skill as a function of time is shown as a solid blue line.

The third aspect of the drift evaluation was to quantitatively assess the simulations using the following metrics:

Separation rate:

$$v_{sep} = \frac{\Delta\left(\vec{x}_{mod} - \vec{x}_{obs}\right)}{\Delta t} \tag{11}$$

Instantaneous skill:

$$Skill_{ins} = 1 - \min\left(\frac{d_i}{S_i}, 1\right) \tag{12}$$

Cumulative skill:

$$Skill_{cum} = 1 - \min\left(\frac{\sum d_i}{\sum D_{obs,i}}, 1\right) \tag{13}$$

Separation rate is defined as the relative rate of change between the modelled drifter position $\vec{x}_{mod}$ and the observed drifter position $\vec{x}_{obs}$. This is a basic metric for trajectory model performance that allows for the simple interpretation of the quality of future forecasts. It is visualized here by plotting the separation between the modelled and observed trajectories as a function of time (Figure 9p–s). The 0FVCOM-based trajectory predictions showed the slowest separation rate, though the results varied based on the drifter type. The FVCOM performed best when predicting the tracks of the CODE/Davis drifters, which suggests that the modelled currents averaged over the upper meter of the ocean correspond best with currents at 0.5 m depth (the approximate centroid of the subsurface portion of the CODE/Davis drifters). Separation rates from SJAP100 and BoF500 were mostly similar, but BoF500 performed slightly better at longer time scales.

Instantaneous skill (adapted from Molcard et al. [49]) aims to answer the question, 'To what extent can the model help us locate a drifting object given a last known position and associated time?'. It defines the model's skill as a function of absolute separation from the initial location at an instant in time, denoted by the ith discrete time interval. In Equation (12), d_i is the separation between the observed and modelled trajectories at a given point in time (denoted by subscript i), and S_i is the linear distance between the observed drifters' position and its deployment location. The normalization by S_i ensures that skill is assigned based on the accuracy of the prediction relative to the magnitude of the observed displacement. For example, a prediction that is within 500 m of the corresponding observation is assigned a higher skill if the observed drifter has travelled 5 km than if it has travelled 1 km. This implies that the model skill can increase with time if Si increases more rapidly than d_i. This behaviour is clearly evident in the results for Sponge drifters, which exhibited a large down-coast displacement. A non-zero instantaneous skill score indicates that the model improves our knowledge of the drifting object's position from the guess that it is still at the last known position. Therefore, all non-zero skill scores indicate that the model is useful, with higher skill scores indicating a narrower margin of error. FVCOM simulations had the highest instantaneous skill score, and perform best when simulating the tracks of the CODE/Davis drifters.

Cumulative skill (taken from Liu and Weisberg, [50]) answers the question, 'How well can the observed trajectory be reproduced when all the aspects of the trajectory evolution are considered?'. It quantifies the skill of the model using cumulative sums to account for the model's ability to reproduce the magnitude, direction, and timing of drift events, and the range of spatial scales in the flow that are encountered by a drifter along its trajectory. In Equation (13), d_i is as defined above, and $D_{obs,i}$ is the distance travelled by the observed drifter during the time i. Cumulative skill is a demanding test of the model's ability to reproduce the timing, magnitude, and direction of drift events, and any skill score above zero was considered an indication of useful model predictions. Figure 9x–aa shows the cumulative skill scores for the prediction of Davis drifter trajectories. Upon close inspection, the FVCOM model slightly outperforms SJAP100 with the solutions indicating non-zero skill for the forecasts up to (and slightly exceeding) 6 h for all drifter types. Again, the FVCOM's increased performance for the CODE/Davis drifters is evident in the cumulative skill scores (Figure 9x–y). We note that both models showed no cumulative skill for >50% of the observed trajectories, however, this might be expected for a highly dynamic region such as SJH where small differences in predictions near the start of the trajectory can lead to substantial deviations at later times.

Interestingly, across all metrics, simulations using BoF500 outperformed the SJAP100 configuration, and according to the cumulative skill scores, BoF500 produced the longest skillful forecast overall—a significant non-zero skill up to 19 h for the CODE/Davis drifters (Figure 9z). This suggests that increased spatial resolution does not necessarily improve the representation of the currents in the uppermost layer of the water column, and a smoothed solution may provide better trajectory predictions. This is related to the unconstrained turbulence in the higher resolution configuration, explained further in Jacobs et al. [51].

3.8. Evaluation of Model Efficiency

The evaluation of model efficiency included a comparison of (1) the number of cores and run-time required to run the models, and (2) the scalability of the models which describes the model's ability to run faster on more cores. The performance of the two models can be directly compared because both were run on the same high-performance computing platform of the Government of Canada. Both the NEMO and FVCOM models for Saint John Harbour met the operational requirement (a 48 h simulation in less than 30 min), but the FVCOM model had much less computational cost overall. For a 48 h simulation, the FVCOM used eight slots (24 cores per slot) with a run-time of 19.5 min; and the three-level nested NEMO model used 31 slots with a run-time of 29.1 min. The high computational cost of NEMO was mainly due to the 100 m configuration (SJAP100). Figure 10 shows that the run time of SJAP100 rapidly decreased when the number of slots increased from 10 to 18, and then gradually decreased with an increasing number of slots. For the FVCOM, the rapid decrease in run time occurred as the number of slots increased from 5 to 8. The scalability curve for the 500 m configuration of the NEMO (BoF500) is closer to that of the FVCOM. The requirements on the number of computer slots are related to the difference in the numbers of computed nodes (grids) among different models: the FVCOM has 56,635 nodes and 108,301 elements, while SJAP100, BoF500 and BoFSS1/36 have 466,718, 161,670 and 39,600 grid points, respectively. This demonstrated a major computational efficiency in using the unstructured grid FVCOM over the structured NEMO for nearshore applications. Despite having a higher resolution in SJH, the FVCOM ran faster than SJAP100.

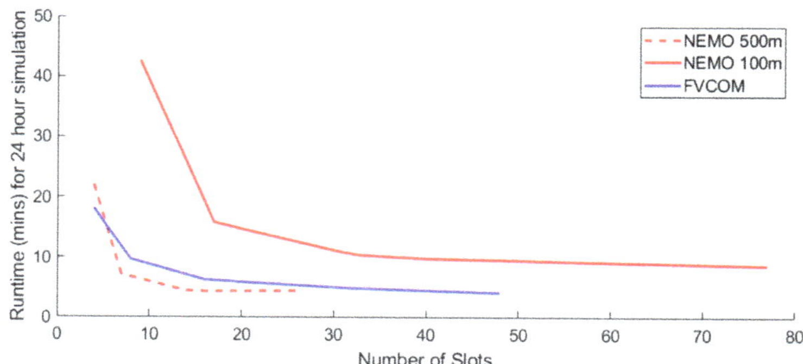

Figure 10. The run time for a 24 h simulation versus the number of slots on the General Purpose Science Computing platform of Shared Services of Canada, with the FVCOM (solid blue curve), the NEMO's 500 m (BoF500, dashed red curve) and 100 m (SJAP100, solid red curve) configurations.

4. Conclusions

This paper represents a significant collaborative effort that took place during the first year of a major Canadian research and development program, the Oceans Protection Plan. The oceanography sub-initiative of the OPP was tasked to develop port-scale models to fit into the multi-scale operational ocean prediction systems in Canada, i.e., to extend the prediction capability from global, basin and coastal scales to nearshore waters. The work included three components: (1) the development of an evaluation process, (2) the development of model configurations with two state-of-the-art open source community models, and (3) the evaluation of the performances of the two model configurations including the comparison of each with the existing operational models. Note that the second and third components were carried out simultaneously, and both model configurations were constantly improved and fine-tuned with the guidance of the evaluation.

The evaluation process includes (1) the selection of the study area, (2) the requirements for the model setup, and (3) the metrics for evaluating the models. The selection of the study area was specific to the nature of the OPP project in that one of the six OPP pilot ports was selected, but the choice of port enabled a general assessment of the models for near-shore port-scale applications. The requirements for the model setup were quite general for evaluating the performance of different source models under the same setting, but by ensuring that both models were configured using the same sources of input data for the model setup and forcing, the differences in the model performances can be mostly attributed to the differences in grid structure, model numerics, and software technology. Thus, the results of this study are valuable for further improvements of the models. The metrics can be used for evaluating any model. They were selected to assess the various aspects of the model results, and collectively, they were able to detect minor contrasts between the two models.

The chosen study area, Saint John Harbour in the Bay of Fundy, features the presence of strong tides, significant river runoff and a narrow tidal-river channel (Reversing Falls). The complicated regional oceanography posed challenges to both the NEMO and FVCOM configurations with the full baroclinic dynamics included. The common challenge was to ensure that the models ran stably, particularly in the narrow channel with strong currents and spatial gradients of temperature and salinity. For this purpose, both models included local treatment of bottom drag parameterization. Previous configurations of the FVCOM model for the Saint John Harbour area did not include atmospheric forcing. Adding atmospheric forcing, particularly the surface heat-flux, was a challenge, but the issue was later attributed to the previously mentioned bug in the code. A challenge for NEMO was the lack of experience in creating a multiple-level nested configuration for nearshore waters. Despite the urgent timeline, both configurations were created and improved during the first year of OPP.

The evaluation metrics were defined based on the existing expertise of the team, through expert consulting and literature research, as well as the available ocean observational data. Evaluation with these metrics led to continuous improvements of both models, that is, errors in model settings (bathymetry, model parameters and forcing) were constantly identified and corrected. The evaluation results presented in Sections 3.4–3.8 were based on the model results obtained toward the end of the first year of OPP. In terms of model accuracy, compared to the observational data, NEMO and the FVCOM achieved comparable metrics for the tidal and non-tidal components of sea level and currents, seasonal variation of the sea surface temperature, vertical profiles of water temperature and salinity, and the trajectories of surface drift. Note that for the evaluation of the drifter trajectories, the impacts of winds were not fully considered. This aspect needs to be included in future work. A major difference identified was that the FVCOM required less computer resources and ran faster than the NEMO. One possible solution to reduce the required number of the computer slots is to reduce the domain size of the finest model configuration (SJAP100), and using the two-way nesting approach to ensure the dynamic interaction across different configurations are properly simulated. Initial testing on two-way nested configuration was achieved after the completion of the evaluation in the first year (results not shown). The results of the model efficiency evaluation suggest that both the NEMO and FVCOM meshes could be further refined, but in doing so, the NEMO model would not meet the operational requirement for runtime due to the limitations of the model numerics and computer efficiency.

Finally, the evaluation of the models was limited by the existing observational data. With increasing resolution, the models are reaching the limits to which the ocean is observed in terms of small scale, rapidly varying, and nonlinear processes. To support the envisaged future of port-scale operational e-navigation systems, we need to reassess what observational capacities are required to support this, both from a monitoring point of view (i.e., real-time and high spatial resolution data), as well as for model development (e.g., delayed-mode data such as moorings).

Author Contributions: Conceptualization, S.N.; methodology, Y.L., G.C.S., N.B.B., P.M., F.D., D.G. and F.J.M.D.; software, S.H., H.B. and G.M.; validation, S.N.; formal analysis, S.H., S.N., H.B.; investigation, S.P.H., J.-P.P., M.O.-S., S.T., H.B., G.M., Y.W., L.Z., X.H., J.C., M.D. and D.G.; resources, Y.L., N.B.B.; data curation, F.P., P.M., F.D. and S.P.H.; writing—original draft preparation, S.N., Y.L., H.B., S.P.H. and G.M.; writing—review and editing, S.H., M.O.-S., S.T., G.C.S., N.B.B., P.M., L.Z., X.H., F.D., D.G. and F.P.; visualization, S.H.; M.O.-S., S.P.H., S.N. and D.G. Supervision, S.N. All authors have read and agreed to the published version of the manuscript.

Funding: This research was funded by the Government of Canada's Ocean Protection Plan.

Acknowledgments: The authors would like to acknowledge the following colleagues for their contributions, guidance and support for this project: Herman Verma, Sarah Scouten, Ji Lei, Adam Drozdowski, Charles Hannah, John Chamberlain, Guoqi Han, Sherry Niven, Kim Housten, Melinda Lontoc-Roy, and Pierre Pellerin. We are grateful to the two anonymous reviewers for the constructive comments that guided the revision of the manuscript.

Conflicts of Interest: The authors declare no conflict of interest.

References

1. Government of Canada. Protecting our Coasts: Oceans Protection Plan. Available online: https://www.tc.gc.ca/en/campaigns/protecting-coasts.html (accessed on 28 May 2020).
2. Government of Canada. CONCEPTS. Available online: http://www.science.gc.ca/eic/site/063.nsf/eng/h_97620.html (accessed on 28 May 2020).
3. Madec, G.; Delecluse, P.; Imbard, M.; Levy, C. *NEMO Ocean Engine*; Note du Pole de modèlisation, IPSL: Paris, France, 2015; ISSN No. 1288-1619.
4. NEMO Community Ocean Model. Available online: https://www.nemo-ocean.eu (accessed on 28 May 2020).
5. Chen, C.; Beardsley, R.C.; Cowles, G. An unstructured grid, finite-volume coastal ocean model (FVCOM) system. *Oceanography* **2006**, *19*, 78–79. [CrossRef]
6. The Unstructured Grid Finite Volume Community Ocean Model. Available online: http://fvcom.smast.umassd.edu/fvcom (accessed on 28 May 2020).
7. Smith, G.C.; Roy, F.; Reszka, M.; Surcel Colan, D.; He, Z.; Deacu, D.; Belanger, J.-M.; Skachko, S.; Liu, Y.; Dupont, F.; et al. Sea ice forecast verification in the Canadian Global Ice Ocean Prediction System. *QJR Meteorol. Soc.* **2015**, *142*, 659–671. [CrossRef]
8. Dupont, F.; Higginson, S.; Bourdallé-Badie, R.; Lu, Y.; Roy, F.; Smith, G.; Lemieux, J.-F.; Garric, G.; Davidson, F. A high-resolution ocean and sea-ice modelling system for the Arctic and North Atlantic oceans. *Geosci. Model. Dev.* **2015**, *8*, 1577–1594. [CrossRef]
9. Dupont, F.; Chittibabu, P.; Fortin, V.; Rao, Y.R.; Lu, Y. Assessment of a NEMO-based hydrodynamic modelling system for the Great Lakes. *Water Qual. Res. J. Can.* **2012**, *46*, 198–214. [CrossRef]
10. Chen, C.; Huang, H.; Beardsley, R.C.; Xu, Q.; Limeburner, R.; Cowles, G.W.; Sun, Y.; Qi, J.; Lin, H. Tidal dynamics in the Gulf of Maine and New England Shelf: An application of FVCOM. *J. Geophys. Res.* **2011**, *116*, C12010. [CrossRef]
11. Wu, Y.; Chaffey, J.; Greenberg, D.A.; Smith, P.C. Environmental impacts caused by tidal power extraction in the upper Bay of Fundy. *Atmos. Ocean* **2016**, *54*, 326–336. [CrossRef]
12. Han, G.; Ma, Z.; DeYoung, B.; Foreman, M.; Chen, N. Seasonal variability of the Labrador Current and shelf circulation off Newfoundland. *Ocean Model.* **2011**, *40*, 199–210. [CrossRef]
13. Foreman, M.G.G.; Stucchi, D.J.; Garver, K.A.; Tuele, D.; Issac, J.; Grime, T.; Guo, M.; Morrision, J. A circulation model for the Discovery Islands, British Columbia. *Atmos. Ocean* **2012**, *50*, 301–306. [CrossRef]
14. Lin, Y.; Fissel, D.B. The ocean circulation of Chatham Sound, British Columbia, Canada: Results from numerical modelling studies using historical datasets. *Atmos. Ocean* **2018**, *56*, 129–151. [CrossRef]
15. Drozdowski, A.; Jiang, D. Modelling internal tides in the Strait of Canso. *Atmos. Ocean* **2020**. [CrossRef]
16. Wu, Y.; Hannah, C.; O'Flaherty-Sproul, M.; MacAulay, P.; Shan, S. A modeling study on tides in the Port of Vancouver. *Anthropoc. Coasts* **2019**, *2*, 101–125. [CrossRef]

17. Church, I.W. Multibeam sonar ray-tracing uncertainty evaluation from a hydrodynamic model in a highly stratified estuary. *Mar. Geod.* **2020**. [CrossRef]
18. Wu, Y.; Chaffey, J.; Greenberg, D.A.; Colbo, K.; Smith, P.C. Tidally-induced sediment transport patterns in the upper Bay of Fundy: A numerical study. *Cont. Shelf Res.* **2011**, *31*, 2041–2053. [CrossRef]
19. Huang, H.; Chen, C.; Cowles, G.W.; Winant, C.D.; Beardsley, R.C.; Hedstrom, K.S.; Haidvogel, D.B. FVCOM validation experiments: Comparisons with ROMS for three idealized barotropic test problems. *J. Geophys. Res.* **2008**, *113*, C07042. [CrossRef]
20. Shchepetkin, A.F.; McWilliams, J.C. The regional oceanic modeling system (ROMS): A split-explicit, free-surface, topography-following-coordinate oceanic model. *Ocean Model.* **2005**, *9*, 347–404. [CrossRef]
21. Trotta, F.; Fenu, E.; Pinardi, N.; Bruciaferri, D.; Giacomelli, L.; Federico, I.; Gi, C. A structured and unstructured grid relocatable ocean platform for forecasting (SURF). *Deep Sea Res. II Top Stud. Oceanogr.* **2016**, *155*, 54–75. [CrossRef]
22. Umgiesser, G.; Melaku Canu, D.; Cucco, A.; Solidoro, C. A finite element model for the Venice Lagoon. Development, setup, calibration and validation. *J. Mar. Syst.* **2004**, *51*, 123–145. [CrossRef]
23. Biastoch, A.; Sein, D.; Durgadoo, J.V.; Wang, Q.; Danilov, S. Simulating the Agulhas system in global ocean models—Nesting vs. multi-resolution unstructured meshes. *Ocean Model.* **2018**, *212*, 117–131. [CrossRef]
24. Wang, Q.; Danilov, S.; Sidorenko, D.; Timmermann, R.; Wekerle, C.; Wang, X.; Jung, T.; Schröter, J. The Finite Element Sea Ice-Ocean Model (FESOM) v.1.4: Formulation of an ocean general circulation model. *Geosci. Model Dev.* **2014**, *7*, 663–693. [CrossRef]
25. Paquin, J.-P.; Lu, Y.; Taylor, S.; Blanken, H.; Marcotte, G.; Hu, X.; Zhai, L.; Higginson, S.; Nudds, S.; Chanut, J.; et al. High-resolution modelling of a coastal harbour in the presence of strong tides and significant river runoff. *Ocean Dyn.* **2019**, *70*, 365–385. [CrossRef]
26. Dupont, F.; Hannah, C.G.; Greenberg, D. Modelling the sea level of the upper Bay of Fundy. *Atmos. Ocean* **2005**, *43*, 33–47. [CrossRef]
27. Shaw, J.; Amos, C.; Greenberg, D.; O'Reilly, C.; Parrott, D.; Patton, E. Catastrophic tidal expansion in the Bay of Fundy, Canada. *Can. J. Earth Sci.* **2010**, *47*, 1079–1091. [CrossRef]
28. Metcalfe, C.D.; Dadswell, M.J.; Gillis, G.F.; Thomas, M.L.H. *Physical, Chemical, and Biological Parameters of the Saint John River Estuary, New Brunswick, Canada*; Technical Report No. 686; Department of the Environmental Fisheries and Marine Service: Halifax, NS, Canada, 1976.
29. Leys, V. 3D flow and sediment transport modelling at the Reversing Falls—Saint John Harbour, New Brunswick. *IEEE Oceans Conf.* **2007**. [CrossRef]
30. Vouk, I.; Murphy, E.; Church, I.; Pilechi, A.; Cornett, A. Three-dimensional modelling of hydrodynamics and thermosaline circulation in the Saint John River Estuary, Canada. In Proceedings of the 38th IAHR World Congress, Panama City, Panama, 1–6 September 2019. [CrossRef]
31. Neu, H.A.; Hydrographic survey of Saint John Harbour, N.B. *National Research Council of Canada, Division of Mechanical Engineering*; Mechanical Engineering Report No. MH-97; National Research Council of Canada: Vancouver, BC, Canada, 1960. [CrossRef]
32. Delpeche, N. Observations of Advection and Turbulent Interfacial Mixing in the Saint John River Estuary, New Brunswick Canada. Master's Thesis, University of New Brunswick, Fredericton, NB, Canada, 2006.
33. Smartatlantic. Available online: https://www.smartatlantic.ca (accessed on 28 May 2020).
34. Drakkar Group. Eddy-permitting ocean circulation hindcasts of past decades. *CLIVAR Exch.* **2007**, *12*, 8–10.
35. Levier, B.; Tréguier, A.; Madec, G.; Garnier, V. *Free Surface and Variable Volume in the NEMO Code*; MERSEA IP report WP09-CNRS-STR03-1A; MERSEA: Paris, France, 2007.
36. Becker, J.; Sandwell, D.; Smith, W.; Braud, J.; Binder, B.; Depner, J.; Fabre, D.; Factor, J.; Ingalls, S.; Kim, S.-H.; et al. Global bathymetry and elevation data at 30 arc second resolution: SRTM30_PLUS. *Mar. Geod.* **2009**, *32*, 355–371. [CrossRef]
37. Flather, R.A. A tidal model of the northwest European continental shelf. *Mem. de la Soc. R. des Sci. de Liège* **1976**, *10*, 141–164.
38. Engedahl, H. Use of the flow relaxation scheme in a three-dimensional baroclinic ocean model with realistic topography. *Tellus A Dyn. Meteorol. Oceanogr.* **1995**, *47*, 365–382. [CrossRef]
39. Milbrandt, J.A.; Bélair, S.; Faucher, M.; Vallée, M.; Carrera, M.L.; Glazer, A. The pan-Canadian high resolution (2.5 km) deterministic prediction system. *Weather Forecast.* **2016**, *31*, 1791–1816. [CrossRef]

J. Mar. Sci. Eng. **2020**, *8*, 484

40. Large, W.G.; Yeager, S.G. *Diurnal to Decadal Global Forcing for Ocean and Sea-Ice Models: The Datasets and Flux Climatologies*; NCAR Technical Note TN-460+ STR; NCAR: Boulder, CO, USA, 2004. [CrossRef]

41. Fairall, C.W.; Bradley, E.F.; Hare, J.E.; Grachev, A.A.; Edson, J.B. Bulk parameterizations of air-sea fluxes: Updates and verification for the COARE Algorithm. *J. Clim.* **2003**, *16*, 571–591. [CrossRef]

42. Smagorinsky, J. General circulation experiments with the primitive equations: I. The basic experiment. *Mon. Weather Rev.* **1963**, *91*, 99–164. [CrossRef]

43. Smagorinsky, J. Some historical remarks on the use of nonlinear viscosities. In *Large Eddy Simulation of Complex Engineering and Geophysical Flows*; Cambridge University Press: Cambridge, UK, 1993; Volume 1, pp. 69–106.

44. Chen, C.; Beardsley, R.C.; Cowles, G.; Qi, J.; Lai, Z.; Gao, G.; Stuebe, D.; Liu, H.; Wu, L.; Lin, H.; et al. *An Unstructured Grid, Finite-Volume Community Ocean Model FVCOM User Manual, v.3.1.6*, 4th ed.; SMAST/UMASSD-13-0701; Marine Ecosystem Dynamics Modelling Laboratory, School for Marine Science and Technology, University of Massachusetts-Dartmouth: New Bedford, MA, USA, 2013.

45. Umlauf, L.; Burchard, H. A generic length-scale equation for geophysical turbulence models. *J. Mar. Res.* **2003**, *61*, 235–265. [CrossRef]

46. Aldridge, J.N. Hydrodynamic model predictions of tidal asymmetry and observed sediment transport paths in Morecambe Bay. *Estuar. Coast. Shelf Sci.* **1997**, *44*, 39–56. [CrossRef]

47. Pawlowicz, R.; Beardsley, B.; Lentz, S. Classical tidal harmonic analysis including error estimates in MATLAB using T_TIDE. *Comput. Geosci.* **2002**, *28*, 929–937. [CrossRef]

48. Marcotte, G.; Bourgouin, P.; Mercier, G.; Gauthier, J.P.; Pellerin, P.; Smith, G.; Brown, C.W. Canadian oil spill modelling suite: An overview. In Proceedings of the 39th AMOP Technical Seminar on Environmental Contamination and Response, Halifax, NS, Canada, 7–9 June 2016; pp. 1026–1034.

49. Molcard, A.; Poulain, P.M.; Forget, P.; Griffa, A.; Barbin, Y.; Gaggelli, J.; Rixen, M. Comparison between VHF radar observations and data from drifter clusters in the Gulf of La Spezia (Mediterranean Sea). *J. Mar. Syst.* **2009**, *78*, 79–89. [CrossRef]

50. Liu, Y.; Weisberg, R.H. Evaluation of trajectory modelling in different dynamic regions using normalized cumulative Lagrangian separation. *J. Geophys. Res.* **2011**, *116*, C09013. [CrossRef]

51. Jacobs, G.A.; D'Addezio, J.M.; Bartels, B.; Spence, P.L. Constrained scales in ocean forecasting. *J. Adv. Space Res.* **2019**. [CrossRef]

Journal of
*Marine Science
and Engineering*

Article

Coastal Flooding and Inundation and Inland Flooding due to Downstream Blocking

Leonard J. Pietrafesa [1,2,*], Hongyuan Zhang [1], Shaowu Bao [1], Paul T. Gayes [1] and Jason O. Hallstrom [3,4]

[1]   Burroughs and Chapin Center for Marine and Wetland Studies, Coastal Carolina University,
     Conway, SC 29528, USA; hzhang@coastal.edu (H.Z.); sbao@coastal.edu (S.B.); ptgayes@coastal.edu (P.T.G.)
[2]   Department of Marine, Earth and Atmospheric Sciences (emeritus), North Carolina State University,
     Raleigh, NC 27695, USA
[3]   Department of Coastal and Marine Systems Science, Coastal Carolina University, Conway, SC 29528, USA;
     jhallstrom@fau.edu
[4]   Institute for Sensing and Embedded Network Engineering, Florida Atlantic University,
     Boca Raton, FL 33461, USA
*   Correspondence: len_pietrafesa@ncsu.edu

Received: 16 May 2019; Accepted: 20 September 2019; Published: 26 September 2019

Abstract: Extreme atmospheric wind and precipitation events have created extensive multiscale coastal, inland, and upland flooding in United States (U.S.) coastal states over recent decades, some of which takes days to hours to develop, while others can take only several tens of minutes and inundate a large area within a short period of time, thus being laterally explosive. However, their existence has not yet been fully recognized, and the fluid dynamics and the wide spectrum of spatial and temporal scales of these types of events are not yet well understood nor have they been mathematically modeled. If present-day outlooks of more frequent and intense precipitation events in the future are accurate, these coastal, inland and upland flood events, such as those due to Hurricanes Joaquin (2015), Matthew (2016), Harvey (2017) and Irma (2017), will continue to increase in the future. However, the question arises as to whether there has been a well-documented example of this kind of coastal, inland and upland flooding in the past? In addition, if so, are any lessons learned for the future? The short answer is "no". Fortunately, there are data from a pair of events, several decades ago—Hurricanes Dennis and Floyd in 1999—that we can turn to for guidance in how the nonlinear, multiscale fluid physics of these types of compound hazard events manifested in the past and what they portend for the future. It is of note that fifty-six lives were lost in coastal North Carolina alone from this pair of storms. In this study, the 1999 rapid coastal and inland flooding event attributed to those two consecutive hurricanes is documented and the series of physical processes and their mechanisms are analyzed. A diagnostic assessment using data and numerical models reveals the physical mechanisms of downstream blocking that occurred.

Keywords: downstream blocking; compound flooding; coastal storm surge and inundation; explosive lateral flooding; hurricane inland and upland flooding

---

1. Introduction

Compound hazards are those events that occur simultaneously or successively whose combination and interaction with underlying conditions amplify the hazardous impacts from individual events [1], creating storm surge and thus seawater inundation. The storm's heavy rainfall, on the other hand, causes surface runoff, sub-surface flow and river flooding. These two flooding processes can have dependence [2] and complex interactions. A higher downstream coastal water level changes the river's downstream boundary conditions, and thus affects upstream river flow dynamics and inland freshwater flooding [3–7]. Simultaneously, the flow of river flow into the ocean can affect coastal sea level changes [8], which, in turn, can act as a feedback effect to further impact the river flooding. Jay et al. [9] and Guo et al. [10] analyzed and documented the relative importance of tides and river flows at different locations along the river channel to coastline at several floodplain wetlands.

An example of a compound hazard occurred in 1999. From late August to early September in 1999, Hurricanes Dennis and Floyd passed along and across the eastern coastal region of North Carolina (NC), and together deposited about 1000 mm of precipitation. The pair created extreme coastal, inland and upland flooding. The combined effects of the hurricanes resulted in massive property damage and led to 74 (56 in NC) human fatalities, due principally to the ensuing flooding. The net cost of the damage ascribed to the flood event was in excess of $6.5B in 1999 and 10.166B in 2019 dollars. The flood event extended from the coast to New Bern NC and Washington NC, well inland and upland. At the time, the flooding inland and upland were not associated with the downstream blocking of the Neuse and Tar-Pamlico Rivers, but rather directly with Floyd's 600 mm of rainfall.

Hildebrand [11] found that from 1887 up to 1999, NC had experienced 83 named tropical storms and 31 hurricanes, but none had resulted in the massive flooding associated with these two 1999 events. However, Pietrafesa et al. [12] further documented the revelation that flooding, instead of winds, was responsible for nominally 65% of hurricane-related property damage in NC. More recently, heavy precipitation events such as hurricanes Joaquin (635 mm rainfall) and Matthew (457 mm rainfall) in South Carolina (SC) in 2015 and 2016, respectively, hurricane Harvey in 2017 in Texas (1828 mm rainfall), and hurricane Florence in 2018 (914 mm rainfall) in NC and SC, are examples of these kinds of heavy precipitation events. Hurricane Matthew resulted in $10.3B in damage in 2016 ($10.92 in 2019) dollars in SC alone. Rivers inland in SC crested to unprecedented levels of 5–6 m over mean water levels (NCEI), such as the Waccamaw River at Freeland and the Congaree River in Columbia, well up to the foothills of the Appalachian Mountains. Following Florence (2018), the Waccamaw River crested at 0.9 m above the level reached during the passage of Matthew (2016). There are wide ranges of spatial and temporal physics scales and thus in reported impacts (p.c. from M. McClam of the South Carolina State Guard), from several kilometers to several tens of kilometers downstream, and hours to days downstream to tens to many hundreds of kilometers upstream and days to weeks and back to tens of minutes upstream. To reduce the risks from these kinds of seemingly hidden and then often explosive flooding events in the future, we need to understand the fluid mechanics of these events that are temporally extensive and spatially massive, often transitioning to short period, laterally explosive from inland to upland and from the coasts to the mountains.

This study discusses the 1999 Dennis-Floyd event and the nonlinear fluid physics that ensued, and presents this case as a precursor to future events in-kind. First proposed by Pietrafesa and Dickey at an Eastern Carolina University (ECU) Hurricane Flood Workshop, in the study reported on below, it is found that in 1999, while Hurricane Floyd was attributed solely for the inland flood damage [13], Hurricane Dennis actually set the stage for the massive inland and upland flooding by changing the downstream boundary conditions. This study documents the events that preceded, were present during, and followed the passages of Dennis and Floyd and offers the possibility of an improved model prediction scheme for inland and upland flooding in coastal states. Moreover, the need to properly initialize the water levels in prognostic numerical models of incoming heavy precipitation events is suggested as both proof of concept and as a warning for the future. The Dennis-Floyd scenario is described in the section to follow, because of the comprehensive data set that is available to study that combined event. A numerical modeling scheme is envisioned and should be developed and employed in the future, if forecasts of coastal, inland and upland flooding are to improve from present-day mathematical architectures.

2. Data and Study Area

The study area, which encompasses eastern NC, and the points of in situ observations, are presented in Figure 1. Data used in this study (cf. Figure 1) include time series of atmospheric winds, precipitation, water levels and water currents. Atmospheric data is from the National Weather Service (NWS) first order stations. River discharge and water level data for Little Washington NC are from the U.S. Geological Survey (USGS). Open ocean coastal sea level and sound-side water level is from the National Ocean Service (NOS). Wind data time series are from the NC Coastal-Marine Automated Network (C-MAN station) located along the coast downstream from Ocracoke Inlet where the + sign is shown, and the Kinston, NC Airport and precipitation data are from the National Weather service (NWS). Sea surface temperature (SST) data are from NOAA's polar orbiting satellite. Sea surface and cloud color data are from the NASA SeaWifs and Infrared Imager satellites. All data, including the hurricane tracks, are available from the National Center for Environmental Information (NCEI) at: https://www.ncei.noaa.gov/.

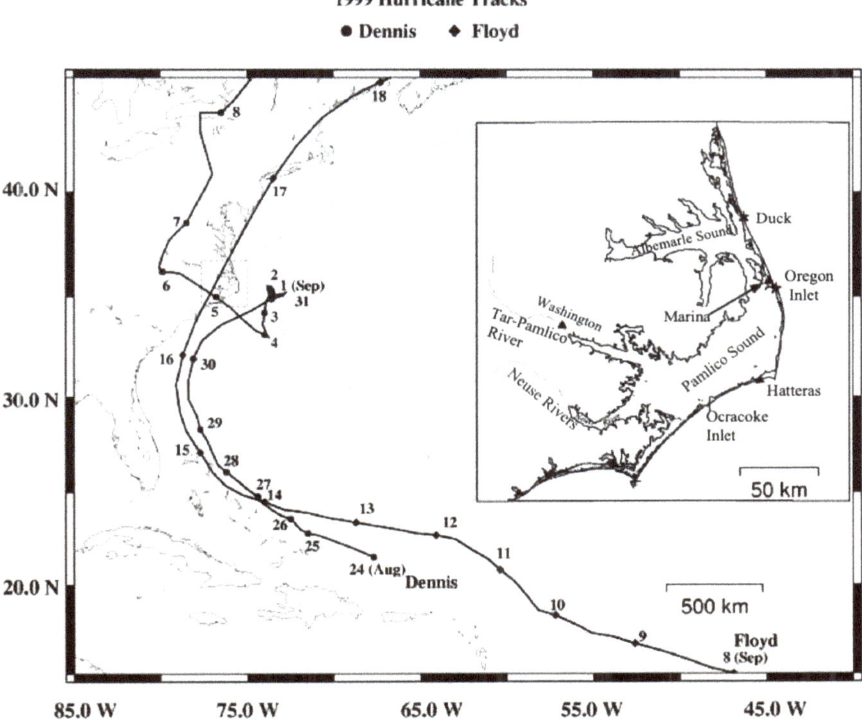

Figure 1. Tracks of 1999 hurricanes Dennis and Floyd. The insert is the eastern NC study area. Triangles represent NWS and USGS data collection sites, and stars represent NOS data collection sites. Data are provided by the NCEI: https://www.ncei.noaa.gov/.

3. Analyses

In Figure 1, we see that Hurricane Dennis entered the region of the NC coast on 30 August and became stationary for 6.5 days off the NC coast east of Cape Hatteras, finally leaving the area on 06 September.

In Figure 2a,e, we see that when the winds blew from the north and northeast, on the western or shoreward side of Dennis' eye, water levels rose within several hours on the open ocean side of the coast at the Duck, NC. The water level (tide gage) data has been low-pass filtered using a 40-hour half power point Lanczos-Cosine filter [14]. Within the sound system, water levels fell in the northeastern end or upper Pamlico Sound, and rose in the upper Tar-Pamlico River, all within three hours. On August 30, water levels rose at Duck by 98 cm, rose at Washington by 21 cm and fell at the Marina by

22 cm. What this indicates is the quick response time of the entire Pamlico Sound to an axial wind, consistent with [15], who reported that water levels respond fully to strong along-sound axial winds within 2 h and 45 min. Thus, water levels in the southwest end of Pamlico Sound set up or rose, and water levels in the northeast end of the sound set down or dropped. The water level at the Marina was out of phase with the water levels at both Duck and Washington over the entire period extending from August 25 to September 20. The rises and falls (Figure 2a–c) and the subsequent differences (Figure 2d) coupled tightly to the NE/SW component (Figure 2e) of the total wind-field (not shown). The difference in water levels was most dramatic from August 30 to September 06, which is coincident with the presence and eventual passage of Dennis. This also had the effect of driving coastal waters towards the three inlets, Oregon, Hatteras and Ocracoke and of driving waters away from them on the Pamlico Sound sides of the inlets. This effected a double suction of water through the inlets, convergence on the outside and divergence on the inside, a process previously reported on by [14]. We show below that this set up in the SW corner had the effect of blocking the flows at the mouths of both the Tar-Pamlico and the Neuse Rivers, so water levels had to rise upstream consequently, filling the water basins to near capacity.

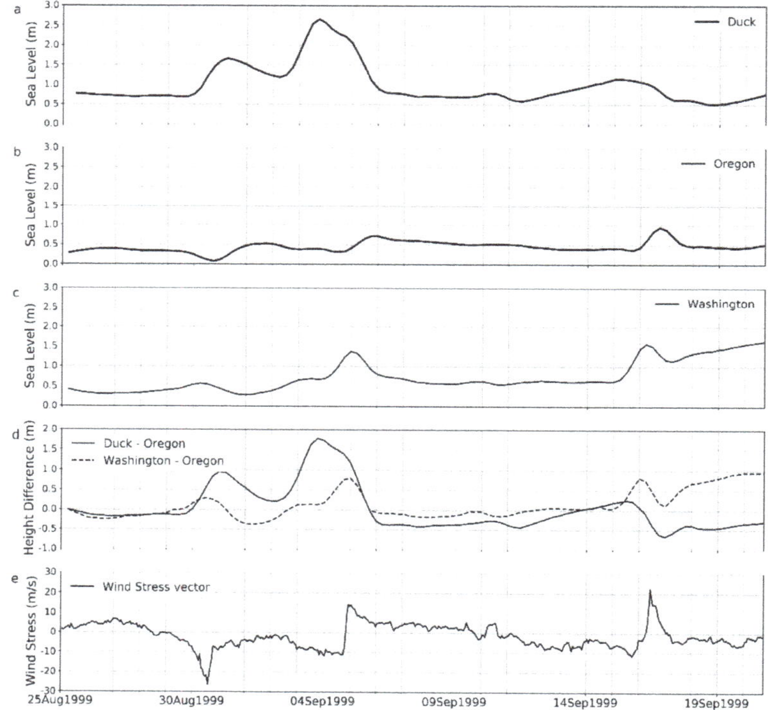

Figure 2. Time series of water level at Duck, NC (**a**), Oregon Inlet (**b**), Washington, NC (**c**), and their difference (**d**). The alongshore wind at Cape Lookout is shown in (**e**) with the positive sign indicating northeastward wind and the negative sign southwestward. Time period is from 08/25 to 09/20, 1999, which encompassed Hurricanes Dennis and Floyd.

Hurricane Dennis then wobbled somewhat off the coast, so at any location the relative wind field changed in intensity with time, while about 280 mm of rain was deposited over the coastal region and the water levels fluctuated. On September 05, Duck water levels peaked at 196 cm higher than they were prior to Dennis' incursion and Washington water levels reached 109 cm higher than they were prior to Dennis' arrival. Inshore Marina water levels were 177 cm below coastal water levels and 85 cm below Washington water levels. Then, as Dennis moved across and finally departed the state on September 06, offshore coastal and upstream river water levels began to return to their prior state.

3.1. Water Transport through the Inlets

The Nichols and Pietrafesa [14] study showed that for periods in excess of a day, the axial flow through Oregon Inlet and sea level slope, are tightly coupled in a 43 cm/sec/meter relationship. Using this stable transform function, we can compute the volumetric flux of water through the inlet. We note that the three inlets from the coastal ocean to Pamlico Sound, Oregon, Hatteras and Ocracoke, are but several km in width, and thus are very spatially narrow. Using 7000 m^2 [14] as the nominal cross section of Oregon Inlet, and an amount in kind for Ocracoke and Hatteras Inlets taken together, we compute the time series of volumetric flux, the cumulative flux of water either in or out of Pamlico Sound is shown in Figure 3. Here we see that following the onset of Dennis, coastal waters began to flood Pamlico Sound via Oregon, Ocracoke and Hatteras Inlets on 30 August and continued doing so until 06 September. During this period, the flux of shelf water into the sound occurred at non-tidal speeds occasionally reaching nearly 1.5 m/s, and the added amount of water that entered Pamlico Sound via Oregon Inlet alone reached the volumetric value of 1.4×10^9 m^3. Ocracoke and Hatteras added another 1.25×10^8 m^3. The salt concentrations of the water masses entering the sound reached 30 parts/1000 or ppt (not shown).

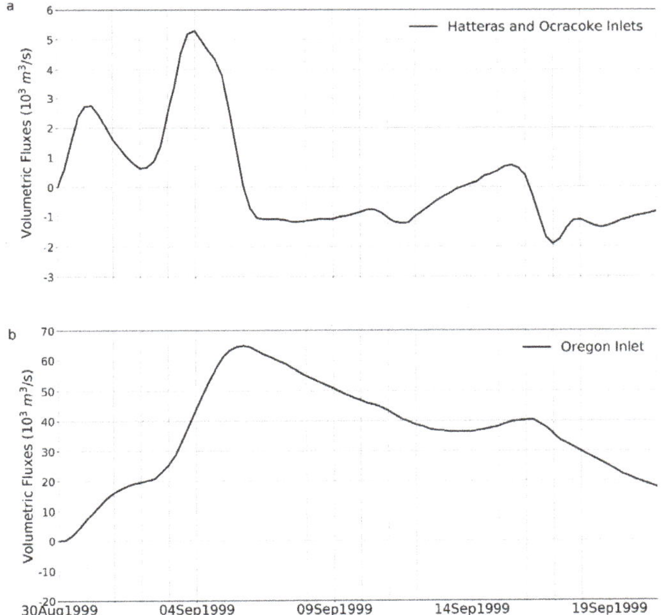

Figure 3. Volumetric Fluxes of water during the passages of Dennis and Floyd (30 August–20 September) through: Hatteras and Ocracoke Inlets (**a**) and Oregon Inlet (**b**). Positive (Negative) is into (out of) Pamlico Sound.

Using the NOS water level records and the estimates of the size of the sound proper [15], we calculate the amount of water that was present in Pamlico Sound at the end of August, prior to the arrival of Dennis, was approximately 1.86×10^9 m^3. Thus, the amount that entered the Sound over the 6.5-day period when Dennis was present increased the total amount of water in Pamlico Sound by 75% to 3.26×10^9 m^3 of water. This additional water flooded the low-lying perimeter of Pamlico Sound.

Following Dennis' departure, the Pamlico Sound system began to drain, through all three barrier island inlets. The total volume of water decreased by about $0.8{\times}10^9$ m^3, but as the M2 tide and weak axial pressure gradient flows were present to drain waters through the inlets (acting as outlets) the sound still retained $2.46{\times}10^9$ m^3 by September 13. However, on September 14 the waters began to rise again as Floyd approached, bringing more precipitation (cf. Figure 4a,b and Figure 2b), along with winds favorable for additional incursions of coastal waters into the sound. By September 16, the additional amount of $0.1{\times}10^9$ m^3 of water added to the system, reached $2.56 {\times}10^9$ m^3 by September 16. This water level was still 38% higher than prior to the arrival of either Dennis or Floyd. Subsequently, the water level gradually fell when the system drained out of the inlets. By September 21, the volume of water in Pamlico Sound proper had declined to $2.25{\times}10^9$ m^3 or 22% more water than was present on 30 August. As can be seen in Figure 1, Hurricane Floyd was present from September 15–17.

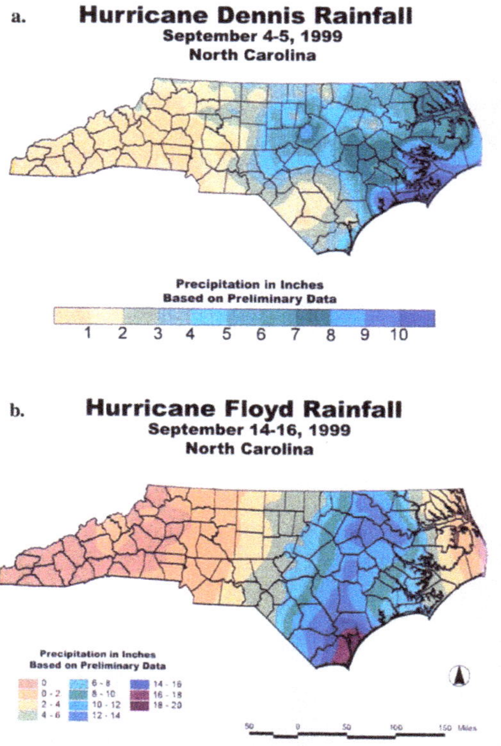

Figure 4. Rainfall bands across NC associated with (**a**) Dennis and (**b**) Floyd in 1999.

3.2. The Water Level Blocking Effect and Subsequent Flooding

The data and the imagery presented show that when Hurricane Dennis sat off the NC coast, it created conditions favorable for the flooding of the Pamlico Sound with coastal waters for 6.5 days. The enormous amount of highly saline coastal water that entered the system added 75% more water to what was already present in Pamlico Sound. Following Dennis' departure, all three sound inlets began to drain using the increased pressure gradient forces from inside to outside the sound along the three narrow inlet axes, as the driving forces. The time series of water levels at the Marina, inside the sound and near Oregon Inlet, and the upper Tar-Pamlico River (Figure 2b,c) and the differences between the two (Figure 2d) offer further evidence for this scenario, which indicates that Pamlico Sound was filled to capacity and thus blocked the mouths of the river-estuary tributaries, specifically the Neuse and Tar-Pamlico Rivers. Then while the system was still backed up, or rather was in a storage mode, and while the region's soils and vegetation were still highly saturated, along came Hurricane Floyd, dropping a then historic record amount of rain over a large swath of NC's coastal and eastern middle interior, inland and upland (Figure 4). The waters in the sound proper were not able to drain to the sea quickly enough as there were no driving forces other than inlet, along river axes pressure gradients. Subsequently the rivers swelled over their banks, drowning the coastal plain and the inshore areas laterally. Moreover, the lateral movement occurred explosively in tens of minutes. The September 23 NASA SeaWifs image of the sound region (Figure 5) strongly suggests that the highly turbid river waters had not yet reached Pamlico Sound proper, though they are clearly present in the rivers.

Figure 5. NASA September 23 SeaWifs satellite image showing surface color following the passage of Hurricane Floyd. Notice the waters in the sound are visually different from those in the Neuse and Tar-Pamlico Rivers.

To assess this possible scenario, we look at the actual daily time series of stream-flow data from both the Neuse and the Tar-Pamlico Rivers. In Figure 6a, following August 30, the flows in both rivers actually dropped until 04 September 04. In Figure 6b, when the flux began to accelerate until September 16–17 when the flows rapidly intensified. By September 19, the total discharge from the two rivers reached the amount of water that entered the sound from offshore during the oceanic flood caused by Dennis. Therefore, Dennis and Floyd acted in concert to cause the extensive flooding. The flooding was so extensive that it reached Greenville, Washington, Kinston and New Bern, NC, cities (not shown) all of the order of tens and hundreds of kilometers inland and upland from Pamlico Sound and the Outer Banks, NC.

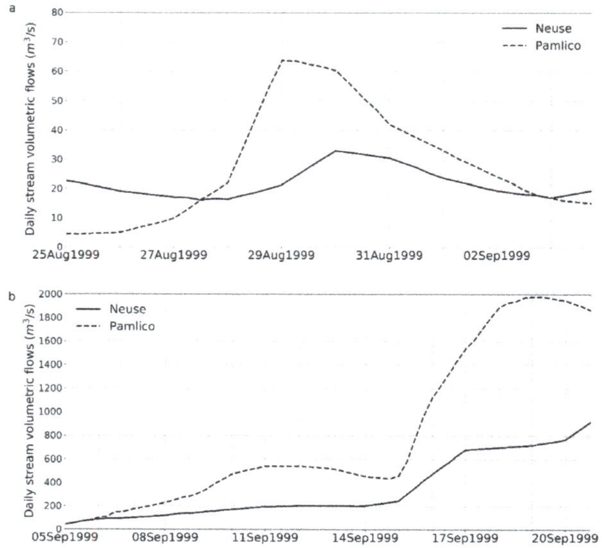

Figure 6. Daily stream volumetric flows in the Tar-Pamlico and Neuse Rivers for the period of (**a**) August 25 to September 04 and (**b**) September 04 to September 20, 1999.

4. Numerical Model Testing

To understand the mechanism of the hydrological process caused by two continuous hurricanes, an idealized one-dimensional hydrological model based on the Saint-Venant (S-V) equations (Equations (1) and (2)), is designed. While a state-of-the science sophisticated three-dimensional model could be applied to reveal the total physics of the compound flooding phenomena in the eastern NC setting, we employ the S-V model to provide simplified, yet revealing foundational physics to the compound flooding phenomena.

$$\frac{\partial A}{\partial t} + \frac{\partial Q}{\partial x} = q_{lat} \tag{1}$$

$$\frac{\partial Q}{\partial t} + \frac{\partial \left(Q^2 / A \right)}{\partial x} + Ag\frac{\partial h}{\partial x} + AgS_f = 0 \tag{2}$$

A is the flow area of cross-section. Q is flow rate. q_{lat} is lateral inflow rate into the channel from rainfall and surface runoff. h is water surface elevation. S_f is friction slope, defined as $S_f = (Q/K)^2$ where K is conveyance from Manning's equation, defined as $K = \frac{1}{n}AR^{2/3}$, where n is Manning's roughness coefficient, R is hydraulic radius R=A/P, P is wetted perimeter. The first two terms of the momentum equation (Equation 2) were ignored; therefore, the momentum equation was simplified as a diffusion wave equation. This implementation of the St. Venant equations is similar to the one used in the WRF-hydro model [16].

$$Ag\frac{\partial h}{\partial x} + AgS_f = 0 \tag{3}$$

The idealized 1-D model employed a 10 km linear river channel with a bed slope of 0.0002. The initial river water depth is set at 1 m. It is assumed that the vertical dimension of the river channel is effectively infinite [16]. Two downstream sea levels were used: 1 m (Control Experiment) as the normal condition, which is the same as the normal river surface height, and 1.5 m (Δ-Sea-Level Experiment), as the condition after the hurricane-induced storm surge. An idealized rainfall amount was added to the river channel. Here we add a lateral discharge with value of 30 m³/s to model one example of the rainfall process. This rainfall has a duration of 30 min. The rainfall line shown in Figure 7 illustrates where the rainfall occurs (3 to 4 km from the river mouth). Results of this model are shown in Figure 7. About half hour after the rainfall was deposited onto and into the river channel, the surface water height in the Δ-Sea-Level Experiment was higher than the Control Experiment up to 2 km from the downstream boundary, and its discharge was slightly less than the Control Experiment 3 h after the rainfall. the higher surface water level intrudes further toward inland reaching up to 4 km from the downstream boundary. The surface water slope in the Δ-Sea-Level Experiment is lower than the Control Experiment, resulting in slower water flow speed (V). However, the higher water surface in Δ-Sea-Level also created a large river cross-section area (A); therefore, the simulated streamflow discharges ($Q = V \times A$) in the Control Experiment and the Δ-Sea-Level Experiment converge after 3 h. Note in this idealized experiment the vertical dimension of the river channel is assumed to be infinite, meaning the increased water can be stored in the river channel. However, in real-world conditions, this increased water will move laterally and become flood waters.

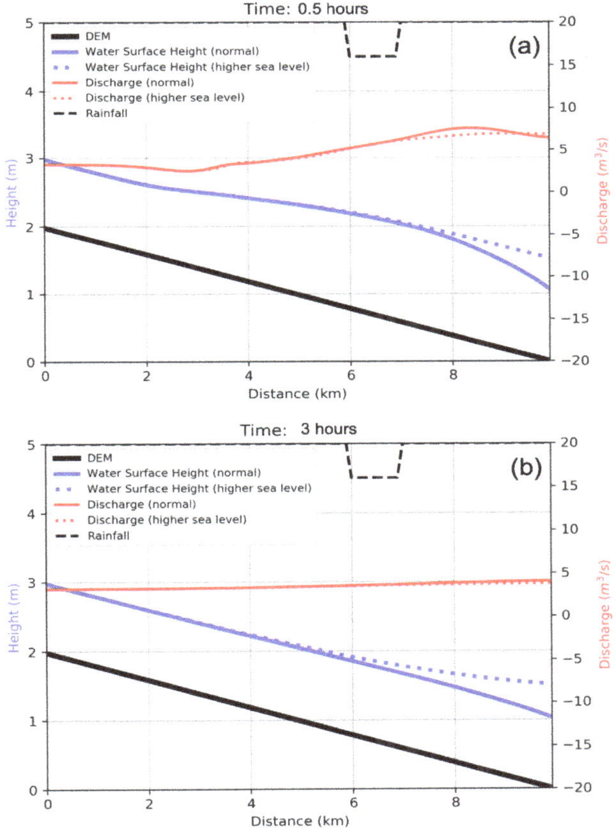

Figure 7. One-dimensional idealized river channel simulation results: (**a**) 30 min after the rain event; and (**b**) 3 h after the rainfall was added to the river channel. The black line is the constant riverbed. The blue lines are the water surface height in the river channel. The red lines denote the streamflow in m³/sec. The bold solid lines show the results from the simulation with normal sea level as the downstream boundary condition, and the dashed lines are those with increased sea level as the initial condition.

Based upon the 1-D idealized numerical experiment, the Princeton Ocean Model (POM), developed by [17] was also used to simulate the above hypothesis. The model incorporates the Mellor–Yamada turbulence scheme, has a free surface to handle tides, sigma vertical coordinates (i.e., terrain-following) to handle complex topographies and shallow regions, a curvilinear grid to better handle coastlines. We employed the POM community model to emulate Pamlico Sound with lateral topography similar to eastern NC and applied it multiple times and by raising the initial water level at the mouths of the two rivers in sequential model runs. In Figure 8a, the initial water level affects the subsequent Hurricane Floyd induced storm surge when the storm surge rises linearly and then becomes highly nonlinear. In addition, in Figure 8b, the initial water level played an even larger role in the lateral flooding of land, with the lateral flooding increases at an increasing rate shown as a nonlinear power law. At zero initial water level, the lateral flooding encompassed about 250 km^2. However, at 2 m above zero, the POM computed lateral flooding rises nonlinearly to 1700 km^2, a 7-fold physics-based nonlinear increase.

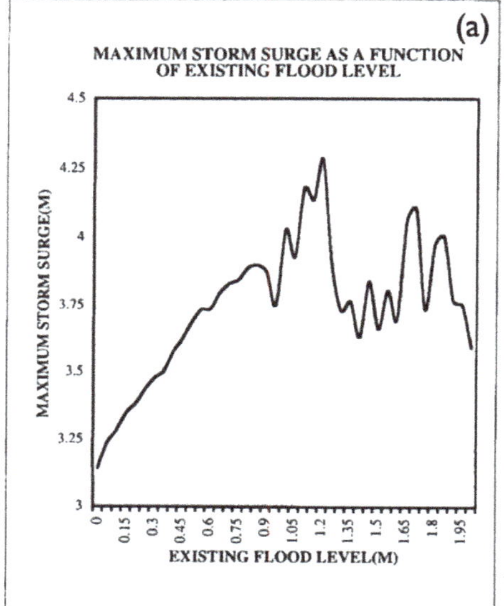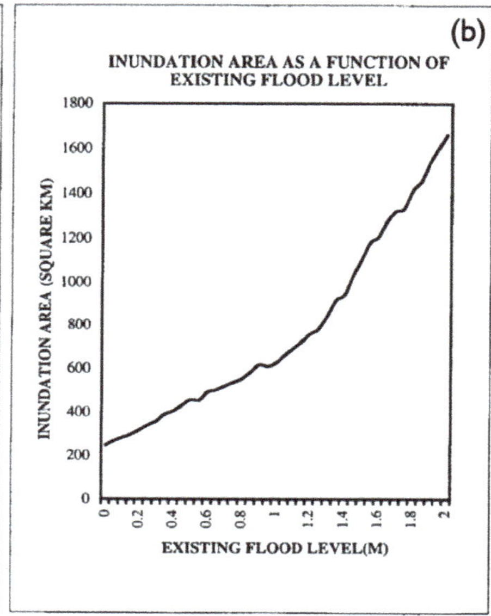

Figure 8. Raising the initial water levels (referred to as the "Existing Flood Level" along the horizontal axis) incrementally for the Pamlico Sound System and then employing the POM model to compute the: (**a**) resulting storm surge; and (**b**) the lateral inundation of the domain.

This simple rendering suggests that when coastal rivers, that is, coastal watersheds, can no longer rise upwards and their carrying capacity to relieve themselves downstream is compromised, additional amounts of rainwater and land runoff will initiate an explosive lateral flooding event. The word 'explosive' is employed because many of the 56 NC casualties occurred on the highways, suggesting that the roads suddenly, within tens of minutes, were inundated with several feet of water and the cars were washed off of the roads and went down the embankments. This phenomenon must be addressed in future diagnostic and prognostic numerical modeling architectures. In fact, the National Water Model (NWM) presently used by the NOAA does not connect to coastal waterbodies on any U.S. coastline, including the Great Lakes. Furthermore, groundwater sources of rainwater can reappear days to weeks following an event or series of events and amplify the unforeseen explosive floods, well upstream, both inland and upland. River water levels may be rising because water is flowing upstream or simply rising in place. Coastal watersheds can transition into storage modes due to downstream blocking; a non-local forcing effect.

5. Discussion, Conclusions and Recommendations

In this study, a 1999 rapid coastal and inland flooding event attributed to two consecutive hurricanes Dennis and Floyd is documented. In summary, the series of physical processes and their mechanisms are the following. (1) Dennis delivered 381 mm of precipitation. (2) Dennis' translational speed slowed to near zero and Dennis hovered off the NC coast for 6.5 days. (3) Dennis' winds mechanically drove coastal waters towards the coast, and within 8 h of onset built up a wall of water along the offshore side of the NC Outer banks coast. These same winds simultaneously drove inshore sound waters from the northeast end of Pamlico Sound towards the southwest end within 3 h. (4) Pamlico Sound was flooded by relatively salty coastal ocean waters. (5) The amount of ocean water which entered the sound system during Dennis' presence was equivalent to 75% of the amount of water already present in Pamlico Sound. (6) The excessive amount of water in Pamlico Sound blocked the flows from the Neuse and Tar-Pamlico rivers, causing the two rivers to go into relative storage modes and thus backed waters up towards the heads of the rivers, filling the watersheds to near capacity, thereby creating the conditions for explosive lateral flooding. (7) Following Dennis' departure, the waters in the sound began to discharge through the three barrier island inlets but before the waters could drain, along came another wet hurricane. (8) Hurricane Floyd deposited a then-record 609 mm rainfall onto already saturated soils. (9) When Floyd arrived, river waters were still blocked at their mouths, and river waters expanded laterally over their banks, thereby flooding the watersheds to record levels many tens to hundreds of kilometers inland and upland. (10) Following Floyd's departure, the entire sound system began to drain through its three barrier island inlets, now functioning as outlets, and continued doing so for several months.

Unfortunately, although the phenomenon of compound flooding has been recognized qualitatively on the conceptual level, the skills of quantitatively predicting such events using numerical models remain poor. The deficiency has been due to the following two reasons. First, traditionally compound events have been treated via a "top-down" perspective, which typically only considers one event and its hazard at a time, potentially leading to an underestimation of risk, as the processes that cause extreme events often interact and are spatially and/or temporally dependent [18]. Such a perspective has led to the fact that in the U.S., flood hazard assessment practices are typically based on univariate methods. Thus, hydrology and oceanography modelers often concentrate only on their own respective domains. For example, procedures for rivers often treat oceanic contributions (e.g., storm surges) using static base flood levels, and do not consider the dynamic effects of coastal water levels. Similarly, flood hazard procedures for coastal storm surge and seawater inundation do not account for terrestrial factors such as river discharge or direct precipitation into urban areas [3,19–21]. Additionally, the National Water Model (NWM), the main tool for the National Weather Service (NWS) to forecast flood events and to issue flood warnings, currently runs on a national domain that cover a majority of the country with gaps along coastal estuaries [22]. Therefore, in order for the coastal and inland flooding attributed to events like hurricanes Dennis (1999) and Floyd (1999) to be predicted in the future, what is required is an atmospheric-oceanic-land-hydrology-hydraulic coupled model system with downstream boundary condition and initial condition properly represented.

The multiscale, multi-physics aspects of this study are considerable and highly informationally provocative. Coastal storm surges are on the order of a meter or more in the vertical and extend alongshore for coastal distances of several kilometers, with a persistence of over several hours to several days. Coastal water level spin-up times are 8 h [23,24]. These wind-driven conditions ride atop semi-diurnal and diurnal tides. Those joint conditions can then ride atop elevated coastal water levels of order kilometers from any prior events, which have left elevated water levels at the mouths of rivers, harbors and estuaries. These conditions then ride atop North Atlantic Ocean Basin seasonal, 3 to 6 months, steric adjustments of a up to 30 cm, over tens to hundreds of kilometers to annually adjusted global sea level rise. All of these are initial conditions and coastal boundary conditions. Then we demonstrated that by elevating initialized sea level at the coast, the storm surge rises but then oscillates in non-stationary and non-linear physics over the periods of hours, while the lateral inundation expands from several kilometers to hundreds of kilometers over collapsing periods down to several tens of minutes. The nonstationary and non-linear fluid multi-scale physics are revealed in this study.

Author Contributions: Individual co-author contributions are: (a) Conceptualization, All; (b) (c) Methodology, All; (c) Software, L.J.P., H.Z., S.B., J.O.H.; (d) Validation, L.J.P., S.B., H.Z., P.T.G.; (e) Formal Analysis, All; (f) Investigation, All; (g) Resources, P.T.G., J.O.H., S.B.; (h) Data Curation, L.J.P., S.B., H.Z.; (i) Writing, Original Draft Preparation, L.J.P., S.B., H.Z., P.T.G.; (j) Writing, Review and Editing, H.Z., S.B., L.J.P.; (k) Visualization, S.B., H.Z., L.J.P.; (l) Supervision, L.J.P., S.B., P.T.G., J.O.H.; Project Administration, P.T.G., J.O.H., S.B.; (l) Funding acquisition, J.O.H., P.T.G., S.B.

Funding: This research was funded by National Science Foundation, grant number CSR 1714015 and CSR 1763294. This work is also supported in part by the Major Research Instrumentation program at the National Science Foundation under award AGS-1624068.

Acknowledgments: The authors wish to thank the U.S. National Science Foundation (NSF) under NSF Grants CSR 1714015 and CSR 1763294, and the Coastal Carolina University Burroughs and Chapin Center for Marine and Wetland Studies for having provided support and facilities necessary to conduct this study.

Conflicts of Interest: The authors declare no conflict of interest.

References

1. Field, C.B.; Barros, V.; Stocker, T.F.; Dahe, Q. *Managing the Risks of Extreme Events and Disasters to Advance Climate Change Adaptation: Special Report of the Intergovernmental Panel on Climate Change*; Cambridge University Press: Cambridge, UK, 2012.
2. Zheng, F.; Westra, S.; Leonard, M.; Sisson, S.A. Modeling dependence between extreme rainfall and storm surge to estimate coastal flooding risk. *Water Resour. Res.* **2014**, *50*, 2050–2071. [CrossRef]
3. Moftakhari, H.R.; Salvadori, G.; AghaKouchak, A.; Sanders, B.F.; Matthew, R.A. Compounding Effects of Sea Level Rise and Fluvial Flooding. *Proc. Natl. Acad. Sci. USA* **2017**, *114*, 9785–9790. [CrossRef] [PubMed]
4. van den Hurk, B.; van Meijgaard, E.; de Valk, P.; van Heeringen, K.J.; Gooijer, J. Analysis of a Compounding Surge and Precipitation Event in the Netherlands. *Environ. Res. Lett.* **2015**, *10*, 035001. [CrossRef]
5. Wahl, T.; Jain, S.; Bender, J.; Meyers, S.D.; Luther, M.E. Increasing Risk of Compound Flooding from Storm Surge and Rainfall for Major US Cities. *Nat. Clim. Chang.* **2015**, *5*, 1093. [CrossRef]
6. Yin, J.; Lin, N.; Yu, D. Coupled Modeling of Storm Surge and Coastal Inundation: A Case Study in New York City during Hurricane Sandy. *Water Resour. Res.* **2016**, *52*, 8685–8699. [CrossRef]
7. Herdman, L.; Erikson, L.; Barnard, P. Storm surge propagation and flooding in small tidal rivers during events of mixed coastal and fluvial influence. *J. Mar. Sci. Eng.* **2018**, *6*, 158. [CrossRef]
8. Piecuch, C.G.; Bittermann, K.; Kemp, A.C.; Ponte, R.M.; Little, C.M.; Engelhart, S.E.; Lentz, S.J. River-Discharge Effects on United States Atlantic and Gulf Coast Sea-Level Changes. *Proc. Natl. Acad. Sci. USA* **2018**, *115*, 7729–7734. [CrossRef] [PubMed]
9. Jay, D.A.; Leffler, K.; Diefenderfer, H.L.; Borde, A.B. Tidal-Fluvial and Estuarine Processes in the Lower Columbia River: I. Along-Channel Water Level Variations, Pacific Ocean to Bonneville Dam. *Estuaries Coasts* **2015**, *38*, 415–433. [CrossRef]
10. Guo, L.; van der Wegen, M.; Jay, D.A.; Matte, P.; Wang, Z.B.; Roelvink, D.; He, Q. River-tide dynamics: Exploration of nonstationary and nonlinear tidal behavior in the Yangtze River estuary. *J. Geophys. Res. Ocean.* **2015**, *120*, 3499–3521. [CrossRef]
11. Hilderbrand, D.C. Risk Assessment of North Carolina Tropical Cyclones (1925–1999). Master's Thesis, North Carolina State University, Raleigh, NC, USA, 2003. Available online: http://www.lib.ncsu.edu/resolver/1840.16/2111 (accessed on 10 March 2019).
12. Pietrafesa, L.J.; Xie, L.; Buckley, E.; Hildebrand, D.; Peng, M.C.; Dickey, D. The Need for a Coastal Estuary/Inland Flood Risk Damage Potential Index. *WIT Trans. Model. Simul.* **2000**, *31*. [CrossRef]
13. NOAA. Hurricane Floyd Floods, September 1999. Service Assessment. 2000. Available online: http://www.nws.noaa.gov/om/assessments/pdfs/floyd.pdf (accessed on 20 April 2019).
14. Nichols, C.R.; Pietrafesa, L.J. *Oregon Inlet: Hydrodynamics, Volumetric Flux and Implications for Larval Fish Transport*; University of North Texas: Denton, TX, USA, 1997.

15. Pietrafesa, L.J.; Janowitz, G.S.; Chao, T.Y.; Weisberg, R.H.; Askari, F.; Noble, E. *The Physical Oceanography of Pamlico Sound*; UNC Sea Grant Publication: Raleigh, NC, USA, 1986.

16. Gochis, D.J.; Yu, W.; Yates, D.N. *The WRF-Hydro Model Technical Description and User's Guide, Version 1.0*; NCAR Technical Document, UCAR: Boulder, CO, USA, 2016; 120p.

17. Blumberg, A.F.; Mellor, G.L. A description of a three-dimensional coastal ocean circulation model. In *Three-dimensional coastal ocean models*; Heaps, N.S., Ed.; American Geophysical Union: Washington, DC, USA, 1987; Volume 4, pp. 1–16.

18. Zscheischler, J.; Westra, S.; Van Den Hurk, B.J.; Seneviratne, S.I.; Ward, P.J.; Pitman, A.; AghaKouchak, A.; Bresch, D.N.; Leonard, M.; Wahl, T.; et al. Future Climate Risk from Compound Events. *Nat. Clim. Chang.* **2018**, *8*, 469–477. [CrossRef]

19. FEMA. Guidance for Flood Risk Analysis and Mapping; Combined Coastal and Riverine Floodplain. Federal Emergency Management Agency. 2015. Available online: https://www.fema.gov/media-library-data/1436989628107-db27783b8a61ebb105ee32064ef16d39/Coastal_Riverine_Guidance_May_2015.pdf (accessed on 15 May 2019).

20. US Geological Survey. *Guidelines for Determining Flood Flow Frequency*; US Geological Survey: Reston, VA, USA, 1981.

21. Zervas, C.E. Extreme Water Levels of the United States 1893–2010. NOAA Technical Report NOS CO-OPS 067; 2013. Available online: https://tidesandcurrents.noaa.gov/publications/NOAA_Technical_Report_NOS_COOPS_067a.pdf (accessed on 10 March 2019).

22. Salas, F.R.; Somos-Valenzuela, M.A.; Dugger, A.; Maidment, D.R.; Gochis, D.J.; David, C.H.; Yu, W.; Ding, D.; Clark, E.P.; Noman, N. Towards Real-Time Continental Scale Streamflow Simulation in Continuous and Discrete Space. *J. Am. Water Resour. Assoc.* **2018**, *54*, 7–27. [CrossRef]

23. Janowitz, G.S.; Pietrafesa, L.J. A model and observations of time dependent upwelling. *J. Phys. Oceanogr.* **1980**, *10*, 1574–1583. [CrossRef]

24. Janowitz, G.S.; Pietrafesa, L.J. Subtidal Frequency Fluctuations in Coastal Sea Level: A Prognostic for Coastal Flooding. *J. Coast. Res.* **1996**, *12*, 79–89.

Journal of
*Marine Science
and Engineering*

Article

Validating an Operational Flood Forecast Model Using Citizen Science in Hampton Roads, VA, USA

Jon Derek Loftis [1,*], Molly Mitchell [1], Daniel Schatt [1], David R. Forrest [2], Harry V. Wang [2], David Mayfield [3] and William A. Stiles [4]

[1] Center for Coastal Resources Management, Virginia Institute of Marine Science, College of William and Mary, Gloucester Point, VA, 23062, USA
[2] Department of Physical Sciences, Virginia Institute of Marine Science, College of William and Mary, Gloucester Point, VA 23062, USA
[3] Tides that Bind LLC, Norfolk, VA 23517, USA
[4] Wetlands Watch, Norfolk, VA 23517, USA
* Correspondence: jdloftis@vims.edu

Received: 1 May 2019; Accepted: 12 July 2019; Published: 26 July 2019

Abstract: Changes in the eustatic sea level have enhanced the impact of inundation events in the coastal zone, ranging in significance from tropical storm surges to pervasive nuisance flooding events. The increased frequency of these inundation events has stimulated the production of interactive web-map tracking tools to cope with changes in our changing coastal environment. Tidewatch Maps, developed by the Virginia Institute of Marine Science (VIMS), is an effective example of an emerging street-level inundation mapping tool. Leveraging the Semi-implicit Cross-scale Hydro-science Integrated System Model (SCHISM) as the engine, Tidewatch operationally disseminates 36-h inundation forecast maps with a 12-h update frequency. SCHISM's storm tide forecasts provide surge guidance for the legacy VIMS Tidewatch Charts sensor-based tidal prediction platform, while simultaneously providing an interactive and operationally functional forecast mapping tool with hourly temporal resolution and a 5 m spatial resolution throughout the coastal plain of Virginia, USA. This manuscript delves into the hydrodynamic modeling and geospatial methods used at VIMS to automate the 36-h street-level flood forecasts currently available via Tidewatch Maps, and the paradigm-altering efforts involved in validating the spatial, vertical, and temporal accuracy of the model.

Keywords: hydrodynamic; modeling; sea level rise; mobile application; app; crowdsourcing; SCHISM; Tidewatch; StormSense; *Catch the King*

---

1. Introduction

Inherently, hydrodynamic models are best validated with water level sensors, due to the precision afforded by defining the timing and depth of inundation at a location in an automated manner [1–4]. As a result of decreased technological costs, low-cost low-energy networks of water level sensors leveraging the Internet of Things (IoT) are beginning to dramatically densify the flood data available in urban environments in coastal areas throughout the world [5,6]. Hampton Roads, VA, USA, hosts one of these IoT networks called StormSense. The network functions as a flooding resiliency partnership between the Virginia Commonwealth Center for Recurrent Flooding Resiliency (CCRFR) and several coastal cities in Hampton Roads [7]. The network's primary goal is to monitor and transmit automated flooding alerts in real time when inundation occurs [8,9] However, an additional function of these sensors is the integration with federal sensor data from the US National Oceanic and Atmospheric Administration (NOAA) and the US Geological Survey (USGS) to validate and improve the Virginia Institute of Marine Science's (VIMS) flood forecast models [7–9].

However, when it comes to logistical considerations and the high cost of maintenance involved in deploying traditional *in situ* or remote water level observing systems, these factors can limit sensor density when even finer scale data are needed, and therefore impede these systems' ability to accurately monitor fine-scale environmental conditions [8,10]. In recent years, the combination of youth who are increasingly globally connected to the internet, and a growing population of retired professionals, poses an opportunity to create a wide-ranging and diverse network of citizen scientists with the capacity to span multiple societal themes [11,12]. Citizen science is public participation in conducting scientific research by non-professional scientists, typically following some form of informal training on data collection. While not a panacea for all inundation monitoring needs, citizen scientists can augment and enhance traditional research and monitoring. Their interest and engagement in flooding resiliency issues can markedly increase spatial and temporal frequency along with an effective duration of sampling. This can reduce time and labor costs, provide hands-on science, technology, engineering, and mathematics (STEM) learning related to real-world issues, and increase their public awareness and support for the scientific process. Naturally, a lack of sufficient professional oversight in citizen science endeavors can introduce caveats to overcome before wide-scale inclusion in an established coastal observing system, yet progress in this underutilized resource is promising [4].

First seeing significant adoption in the US in the aftermath of 2012 Hurricane Sandy, citizen science flood monitoring efforts first became useful through mobile phone pictures capturing inundation with a time-indexing landmark in view, such as a clock tower or local bank clock [13,14]. These pictures gave credence to the digital medium with the advent of enhanced-GPS, which leverages the Global Positioning System's (GPS) satellite constellation along with nearby cell towers to better triangulate a user's position on the ground. These tools have now begun to rival the utility of government-sponsored post-event flood monitoring efforts, such as the USGS' high water marks [15]. While the latter approach affords confidence for model validation through a trusted agency for superior accuracy, the former possesses a greater capacity to document everywhere flooding occurs, with the inherent risk of potentially less accurate validation data. Regardless, collection of data at the local scale in public spaces where flooding is prevalent, such as streets, public right-of-way access spaces, and parks, can improve model prediction by properly resolving flow around small-scale features in the built environment [12]. Additionally, the model's predictive acumen can be enhanced via improved calibration of assumptions, such as: (1) Better friction parameterization of different land cover types, (2) improved aerial elevation estimates of occluded roadway overpasses, and (3) identification of tidally-susceptible subterranean drainage infrastructure junctions (where tidal waters can enter city streets several blocks from the water's edge). Thus, quality assurance of flood validation data near these fine-scale features can become valuable model improvement assets through the proper training of a citizen scientist network [16].

Through technological progression, many effective methods for mapping inundation and flood depths have been developed using GPS, photo tagging, Augmented Reality (AR) image landmark recognition, and Quick Response (QR) codes [7,11,13]. Naturally, the emergence and growing necessity of smart phones in modern living has popularized the prominence of these recording methods. Additionally, the ease of access afforded by mobile applications for making insurance claims, verifying flooding for municipal government attention, and greater scientific aspirations has increased the intrinsic value of personal flood mapping [17,18]. Thus, flood-observing mobile applications, like "MyCoast" [19] and "Sea Level Rise" [12], or crowdsourcing web data geo-forms, like those implemented at the state [20,21], country [22,23], and international level [24], have emerged for myriad resiliency purposes. Typically, these applications exist to verify claims of flooding, validate flood forecast models, or inform long-term flood planning efforts [19–24]. Mobile flood mapping platforms and applications have recently become information repositories that provide a living data archive of flood observation data with sufficient recording frequency and data density in urban areas where flooding is prevalent [25]. However, these tools have been shown to be of less utility in rural coastal areas, where statistically, less people are present and motivated to vigilantly monitor inundation, and where enhanced-GPS signal strength is diminished due to less reliable cellular broadband coverage [26]. Yet, over time, these data sets can even become their own autonomous data-driven flood prediction models via sea level trend extrapolation when combined with Digital Elevation Models (DEM) [16]. Thus, high-resolution street-scale hydrodynamic models have recently found a new way to validate their predictions, and a cost-effective method for correcting erroneous elevation assumptions from aerial lidar surveys. This includes occluded areas in heavily canopied flood-prone areas and built infrastructure, such as box culverts, highway overpasses, and bridges that impact proper hydrologic drainage in flooding conditions [27].

A proactive and safe way to leverage these technological advances in citizen science flood monitoring methods without waiting for a major storm to elucidate inaccurate model assumptions is to map the incidence of "nuisance flooding." This approach takes advantage of mapping inundation in places where it frequently occurs with minimal danger to the reporter, and can identify issues with modeled flood forecasts without waiting for a major tropical or extratropical storm event to identify them first [12]. Hampton Roads, VA, USA experiences tidal nuisance flooding 12 to 18 times a year [28]. This is a frequency that amounts to no less than one cumulative week per year that low-lying streets in the region are inundated [29,30]. This chronic flooding fatigue can make it easy to forget that intermittent tidal flooding events cost cities and their residents time and money [30,31]. Of these tidal inundation events, the highest astronomical tide of the year has become known as the king tide [20]. While not a scientific term, a king tide is a name that refers to an exceptionally high tide, without the consideration of atmospheric amplification from wind or waves [21,23]. These predictable king tide events can be estimated far in advance and make coordinating and mobilizing a volunteer effort to track their inundation extent easy, while maximizing the opportunity for local weather impacts to potentially amplify the inundation observed [20].

This manuscript describes methods employed at VIMS to disseminate automated inundation forecasts called Tidewatch Maps. The forecasts function as an operational flood forecast model, which leverages the open-source Semi-implicit Cross-scale Hydro-science Integrated System Model (SCHISM) to automatically compute storm tide simulations throughout the entire US East and Gulf Coasts (Figure 1). SCHISM then translates those water level outputs to 220 localized lidar-derived sub-grid modeled sub-basins ranging from 1 to 5 m resolution to calculate 36 1-h geospatial flood depth layers covering all of Tidewater Virginia. SCHISM and the Tidewatch Map currently download inputs and update mapping outputs twice daily, every 12 h. Reliable inundation prediction depends upon accurate simulation of large-scale inundation of the tidal long wave during a king tide to successfully propagate from the ocean, through the continental shelf, estuarine systems, into creeks, and ultimately city streets, and rigorous conservation of fluid momentum and mass as flood waters permeate the built environment. These Tidewatch forecast maps were benchmarked in Hampton Roads by >100,000 GPS-reported high water marks collected by citizen scientists during two king tide flooding events occurring in 2017 and 2018.

The following sections highlight how coastal communities are being meaningfully engaged in coastal ocean observing mechanisms and the research efforts they support. What follows is a description of: (1) A citizen science flood mapping project called *Catch the King* based in Hampton Roads, VA, (2) effective volunteer training methods using cell phones to provide meaningful GPS observations for effective model validation, (3) hydrodynamic modeling approaches used for expediently simulating and publicly mapping near-term inundation, along with (4) a summary of the results. The paper concludes with an identification of the modeling and monitoring challenges and potential solutions for modeling and citizen science efforts in the future.

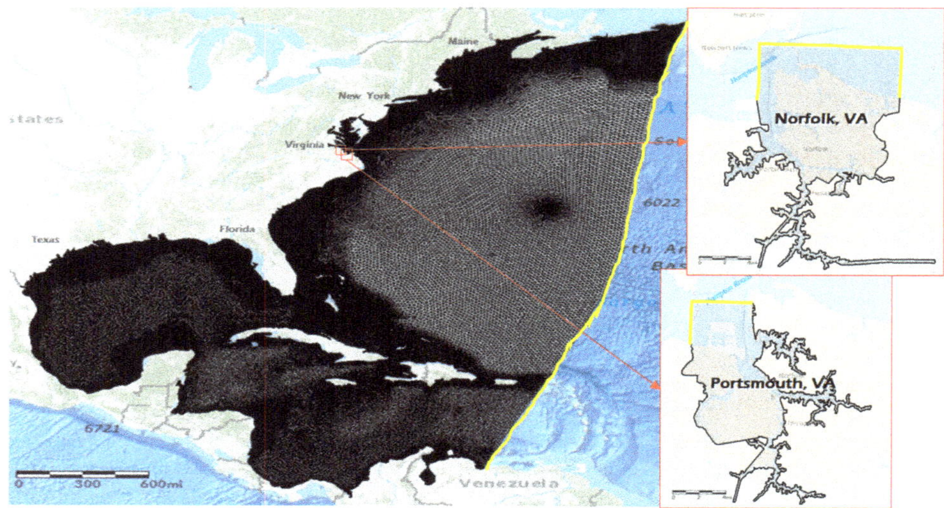

Figure 1. The SCHISM Hydrodynamic model grid used to drive the Tidewatch storm tide prediction model, and inset at right, are some example city-scale sub-grid model sub-basins for Norfolk (top) and Portsmouth (bottom) driven by SCHISM using Dirichlet boundary conditions.

2. Methods

2.1. Citizen Science Flood Monitoring Data Collection

Catch the King is a citizen science GPS data collection effort centered in Hampton Roads, VA, which aims to map the king tide's maximum inundation extents, with the goal of validating and improving predictive models for future forecasting of increasingly pervasive nuisance flooding [16]. The aptly-named effort is the world's largest simultaneous citizen science GPS flood data collection effort, and it is aimed at benchmarking the highest astronomical tide of the year, the king tide. Certified by Guinness World Records for having 'the most contributions to an environmental survey' on the planet, *Catch the King* was effectively publicized and promoted by the local news media, in Hampton Roads, VA [27,32]. High citizen engagement during a king tide inundation event resulted in an average of 572 GPS-reported high water marks being recorded per minute during the hour surrounding the king tide's peak [12]. Time-stamped GPS flood extent measurements and photographic evidence were reported via the free 'Sea Level Rise' mobile application, coinciding with the king tide, observed at 13:32 UTC (Universal Time, Coordinated) on 5 November 2017, in Hampton Roads, VA. Ultimately, *Catch the King* surveyed a total of 59,718 high water marks and 1582 photographs through 722 individual volunteers in its inaugural year [32,33].

Following 2017's success, *Catch the King* was repeated during the highest astronomical tide of 2018 on 27 October. This event was less attended as it immediately followed a strong nor'easter in a day prior, on 26 October. Still, significant response from the event's many volunteers, fueled by the local media partners' coverage leading up to the event, and 42 separate volunteer training events held all over Hampton Roads, resulted in 347 participants collecting 33,847 time-stamped GPS maximum flooding extent measurements and 458 geotagged photographs during the event [34]. There were 141 additional volunteers who collected 3881 GPS data points during the nor'easter's peak flood period coinciding with the high tide the night before the king tide. Combined, those totals are 488 participants and 37,728 GPS points during the 2018 *Catch the King* and nor'easter event, with 431 of those being unique volunteers and 57 having mapped the king tide on both the 26 and 27 October 2018 [35].

As the sea level rise and tidal flooding increasingly impact coastal Virginia, *Catch the King* offers residents a chance to crowdsource vital information about the tides' reach. Thus, time-stamped GPS data points and photographs were collected by volunteers to effectively breadcrumb/trace the high water line by pressing the 'Save Data' button in the 'Sea Level Rise' mobile app every few steps along the water's edge during the king tide's peak in 2017 and again in 2018. Initial responses collected by the free 'Sea Level Rise' mobile application were relayed to the event's dedicated volunteers via the local news media event's volunteer coordination Facebook page. In the days following each event, several people provided additional GPS data points they collected through Esri's Geographic Information System (GIS) Collector App and crowdsourcing geo-forms through ArcGIS Online that were missing from the initial cited statistics (Figure 2). This occurred as a few volunteers noticed their data initially missing from the interactive map as it was published in the Daily Press and The Virginian-Pilot among other media sources, and were later included in the final event totals and added to the final data map [27]. These updated totals were ultimately reported back to the community a month after each inundation event in 2017 and 2018 through a volunteer thank you and data review event [16,35].

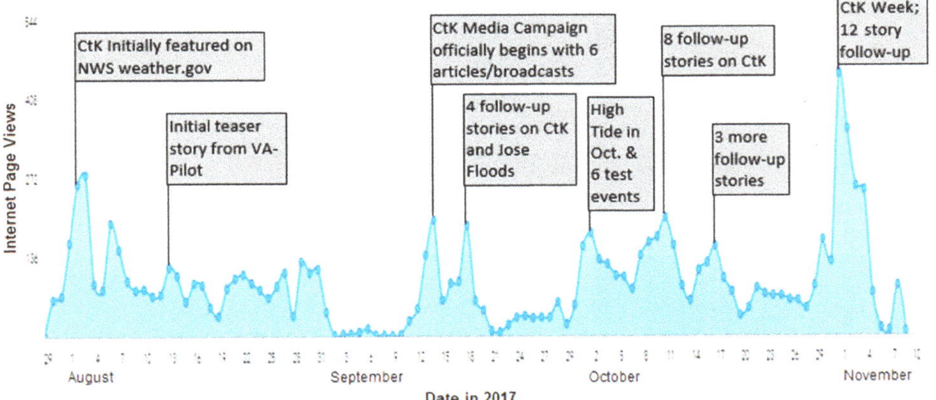

Figure 2. Catch the King (CtK) citizen engagement time series chart corresponding to major news releases used to garner successful volunteer training participation and support leading up to the king tide on November 5, from July 30–November 10, 2017. According to ArcGIS Online's data metrics, the volunteer invitation story map received 10,137 page views, in little over 3 months for an average of 93 page views per day. Story Map at: http://arcg.is/1f8W1q.

2.2. Volunteer Coordination for Flood Extent Validation

The volunteer coordination effort involved a hierarchical scheme led by an adept volunteer coordinator, Qaren Jacklich, from the Chesapeake Bay Foundation who has successfully led the 'Clean the Bay Day' litter and debris removal initiative in Chesapeake Bay for several years prior to *Catch the King*. Below the volunteer coordinator were over 120 volunteer "Tide Captains," who led smaller organized groups of volunteers. In many cases, these tide captains were knowledgeable school teachers, volunteer organization leaders, and enthusiastic users of the Sea Level Rise mobile app, who trained neighbors, friends, and family in their community over a series of separate volunteer training events held all over Hampton Roads, VA, USA [16]. Citizen scientists were trained in the use of the 'Sea Level Rise' mobile application to capture three types of flood data useful for model validation:

(1)   GPS locations for mapping high water contours during flooding (Figure 3);

(2)   Taking field notes for important observations or denoting errata; and

(3)   Uploading time-stamped, geo-tagged pictures, including directional facing information.

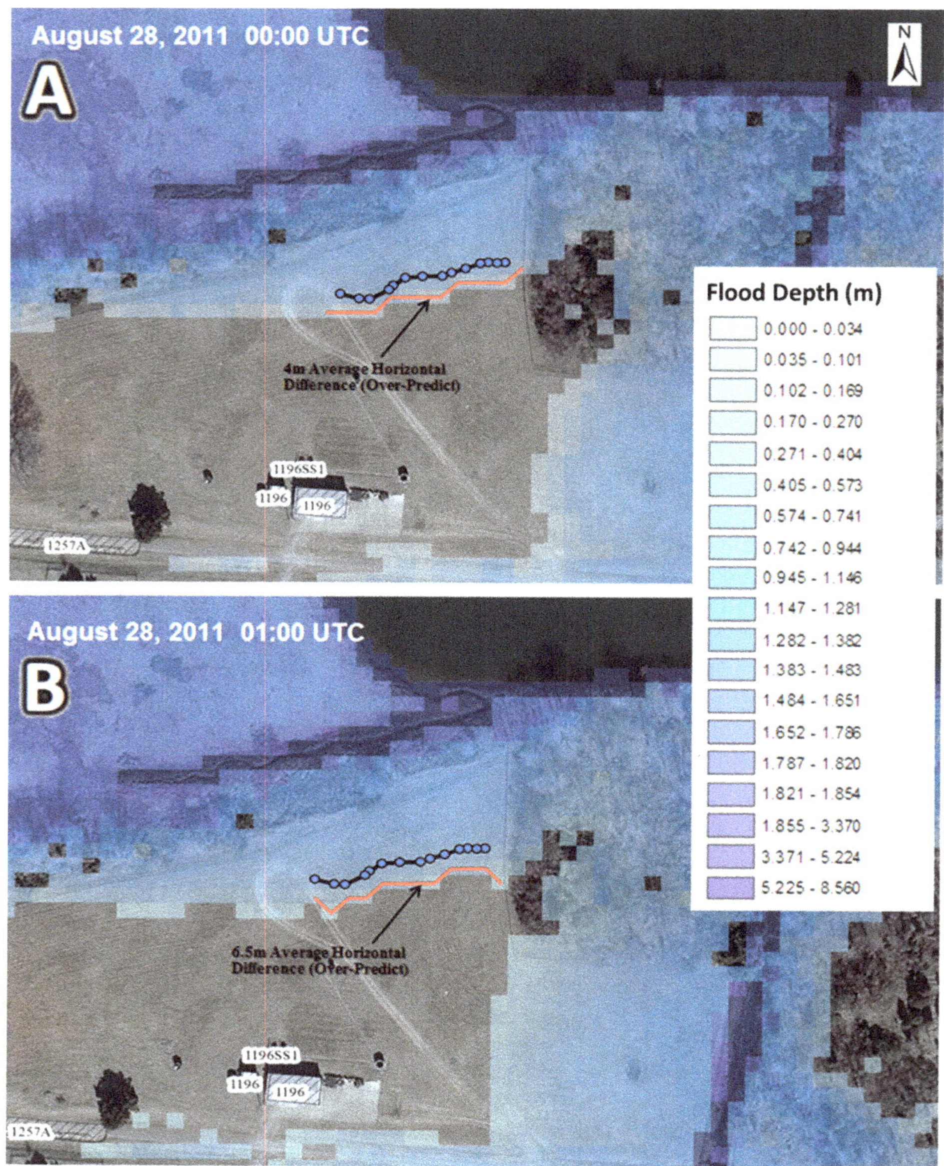

Figure 3. Comparison of GPS maximum flood extent observations (depicted as blue dots) in Hampton, VA, following debris lines remaining after Hurricane Irene 2011, used to geospatially verify model performance at (**A**) 00:00 UTC and (**B**) 01:00 UTC on 28 August 2011. GPS data collected by volunteers effectively illustrate the challenge with temporal matching for model comparison.

Training volunteers in proper, uniform data collection and appropriate geospatial filtering for these three data types is critical to recording an accurate location history for mapping tidal flooding and ensure the most effective approach for collecting error-free data for model validation. More than 45% of the volunteers were directly trained in a training session in 2017's event, while nearly 60%

attended a training in 2017 [35]. However, in both years, many volunteers were still able to register as users and meaningfully map flooding in their community through Catch the King without any formal training, partially owing to the relatively intuitive interface of the app.

The 'Sea Level Rise' mobile app streamlined data collection and made model validation easier through a hierarchical internal quality assurance mechanism. This allowed the Catch the King event coordinator to limit participation to certain trusted registered users, and filter data permissions, such as photo uploads and GPS data collection, only to certain trained users (labeled as 'Champions' in the app interface). Post-event, tide captains and the volunteer coordinator could download their event data as csv files after the specified time window for their flood monitoring event expired, and even retroactively flag volunteer's data where consistently erroneous data points were measured. The resulting maps shown in the results and discussion sections present these dense data maps surveyed during the 2017 and 2018 king tide inundation events, and present the mean horizontal distance difference (MHDD) comparative spatial calculations between the modeled maximum flood extent contours and citizen science flood validation data sets for each king tide flood event, followed by the lessons learned.

2.3. Tidewatch Storm Tide Modeling

Each year, prior to the king tide flood event, researchers at VIMS and the CCRFR design a web map to direct volunteers to public places that are forecasted to flood during the King Tide using VIMS' hydrodynamic models. This method proved impactful, as the 2017 *Catch the King* volunteer recruitment story map reached over 10,000 page views before the king tide in less than 3 months (Figure 2). Thus, the story map effectively conveyed the value of inundation forecasting by showcasing flooding impacts during the last major storm event in the region, 2011 Hurricane Irene, and the importance of time-stamped GPS data for tidal calibration and event calibration of models for improvement purposes (Figure 3). This was valuable to visually explain the value of accurate data collection both temporally and spatially to the citizen scientists.

As Figure 3A shows, on 28 August 2011 at 00:00 UTC after Hurricane Irene, 30 min before the storm surge peak was observed at the nearest sensor, the 14 GPS points that comprise this maximum extent contour compare well with an MHDD of 4 m to the nearest model grid cell center point. However, the model's prediction from an hour later, 30 min after the storm surge peak was observed at the nearest sensor shown in Figure 3B, the maximum flood extent compares less favorably, changing the MHDD to 6.5 m, and illustrating the burden of timing for reliable model comparison. Thus, during the king tides in 2017 and 2018, GPS data points were collected by each year's event's many volunteers to breadcrumb maximum inundation extents in public spaces and time-stamped (Figure 4). This approach was used to coordinate and validate the flooding extents across 17 coastal cities and counties in Virginia, USA, by enlisting the aid of over 1000 volunteers for approximately an hour once a year to walk outside and press the 'Save Data' button in the 'Sea Level Rise' app every few steps along the water's edge near them during the king tide.

The approach to presenting time series information and inundation areas for a flood model at the street-level can be a difficult task for development and comprehension. To simplify the approach, the open-source SCHISM model was developed at VIMS and used to compute Tidewatch's temporal-spatial inundation maps [36]. SCHISM is an open-source community-supported modeling system, designed for the effective simulation of 3D baroclinic circulation across ocean-to-creek scales. The model incorporates a wide range of physical and biological processes in a comprehensive modeling system that has been validated in many world-wide applications, ranging from general circulation [37], tsunami inundation [38], storm surge inundation [39], ecology [40], sediment transport [41], and oil spills [42]. The model is uniquely capable of accurately representing physical structures (both nature-based and engineered) in an inundated area in the model computations, not simply in the output displays. Furthermore, the outputs from this model can be nested with other hydrodynamic grids to provide street-level (1–5 m scale) urban inundation predictions for individual land parcels [43]. The results may be presented as high-resolution movies of flooding scenarios over multi-day periods, including tidal cycle variations, or translated and converted into GIS mapping applications, as seen here with Tidewatch for added utility.

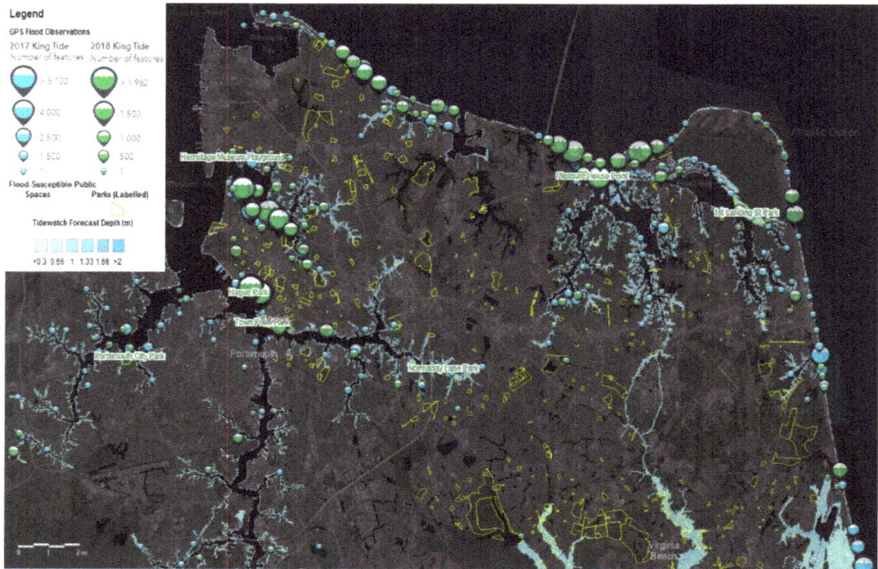

Figure 4. Comparison map of *Catch the King* citizen science king tide data collection between 2017 (blue dots) and 2018 (green dots); points are aggregated by density. Yellow polygons illustrate a public spaces inventory, where volunteers were encouraged to map, while green-labelled park polygons were places where official *Catch the King* volunteer training events occurred. There were 35 such training events in 2017 and 43 in 2018: http://arcg.is/1vK8ru.

SCHISM is used to drive VIMS' Tidewatch's storm tide inundation maps, and the automated workflow to accomplish this is sub-divided into three tasks: (1) Preprocessing of the model grid and retrieval of the hydrodynamic model inputs, (2) SCHISM model simulation, and (3) post-processing retrieval of SCHISM inundation model outputs for GIS mapping, as illustrated in Figure 5:

(1)   Retrieval of SCHISM Hydrodynamic Model Inputs

        a.   Atmospheric data: The SCHISM hydrodynamic model is used to automate storm tide simulations based upon atmospheric wind and air pressure data available from the 05:00 UTC and 17:00 UTC updates of the US National Weather Service's NAM-nest 5 km atmospheric forecast model. For the 2017 and 2018 king tide flooding events, the 17:00 UTC, atmospheric forecast update from the previous night was used.

        b.   Flux boundaries: Riverine boundaries defined at waterfalls and key discharge points at the fall lines of Chesapeake Bay's major estuaries were driven by flow rates predicted by the national water model and obtained at intervals similar to the atmospheric data.

        c.   Open boundaries: Inputs for tidal open boundary grid nodes were harmonically computed and estimated for the amplitude, phase, and frequency for 16 tidal constituents.

(2)   2D SCHISM Model Simulation

        a.   Initiate simulation after archiving the previous run, and complete a successful check for all input files.

b. Hydrodynamic model simulation along the US East Coast and Gulf of Mexico used a geometric model mesh consisting of 2,348,351 nodes and 1,565,567 elements (Figure 1).

c. Post-processing to combine multi-core parallel outputs into binary model outputs, and then into ASCII text for geospatial translation via Python scripts using the raster calculator in GRASS GIS.

d. Extraction of results at Tidewatch charts stations for surge guidance and key points for Tidewatch Maps at 220 Hydrologic Unit Cells (HUCs) throughout tidewater Virginia.

(3) Retrieval of SCHISM inundation model outputs for GIS post-processing

a. Extraction of source data from the station output from the 2D hydrodynamic forecast simulation.

b. Clear the previous run's data to archive, and import new simulation data to differentiate each day's morning and evening simulation updates,

c. Construct a web-enabled time-aware street-level GIS map from SCHISM grid outputs.

d. Publish the output map of 37 time-aware rasters overlaid with Tidewatch Station time series data.

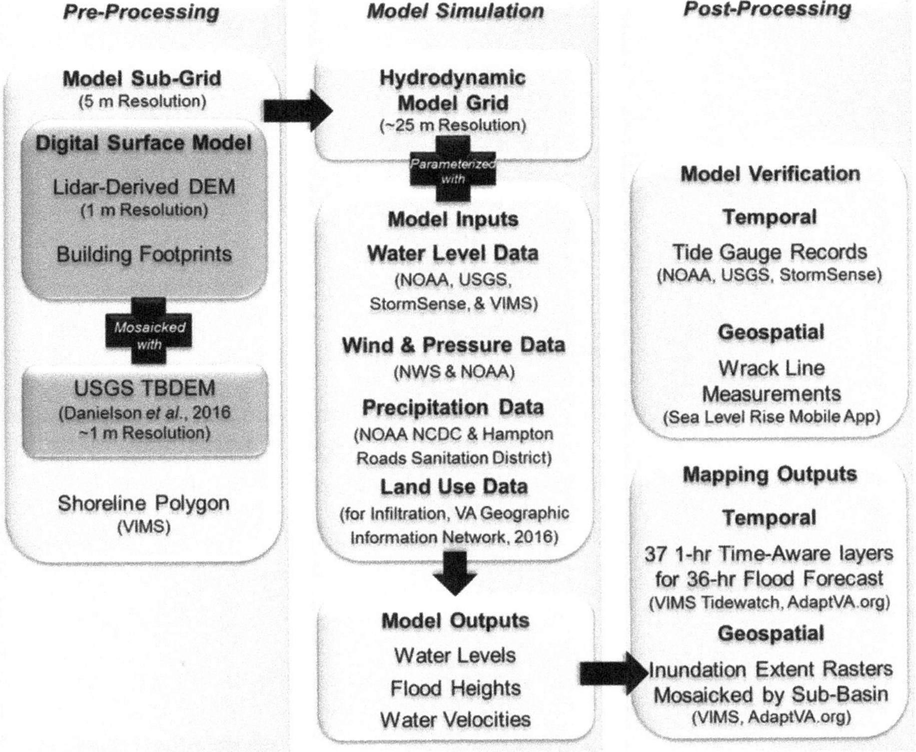

Figure 5. Automation flowchart for SCHISM model initiation and simulation for every 36-h forecast. The Tidewatch model updates twice daily, outputting mapping outputs at AdaptVA.org at noon and midnight local time (05:00 and 17:00 UTC).

Upon completing these steps, model and visualization performance metrics indicated, on average, that step one requires 1 h to complete pre-processing, step two requires 1.5 to 2.5 h to run the simulation (variable based upon volume of inputs), using 72 nodes on VIMS' high performance computing cluster, and step three requires 2 to 3 h in post-processing to combine the model's binary outputs, and translate/index them to hourly geospatial outputs to a convenient interactive web-map presentation format. The SCHISM model consumes approximately 2.40 GB of input data to model the US East and Gulf Coasts for every simulation, twice daily (1.75 TB annually). Comparatively, the SCHISM model outputs the mosaicked results of over 200 combined sub-grid sub-basin rasters to form the Tidewatch Map's composite 36 h. time aware layer cache at 16 zoom-level pyramids for a 21.25 GB product after each simulation (15.48 TB annually). These steps allowed users to interact with flood data from the global scale to street-level in a single web map (Figure 6).

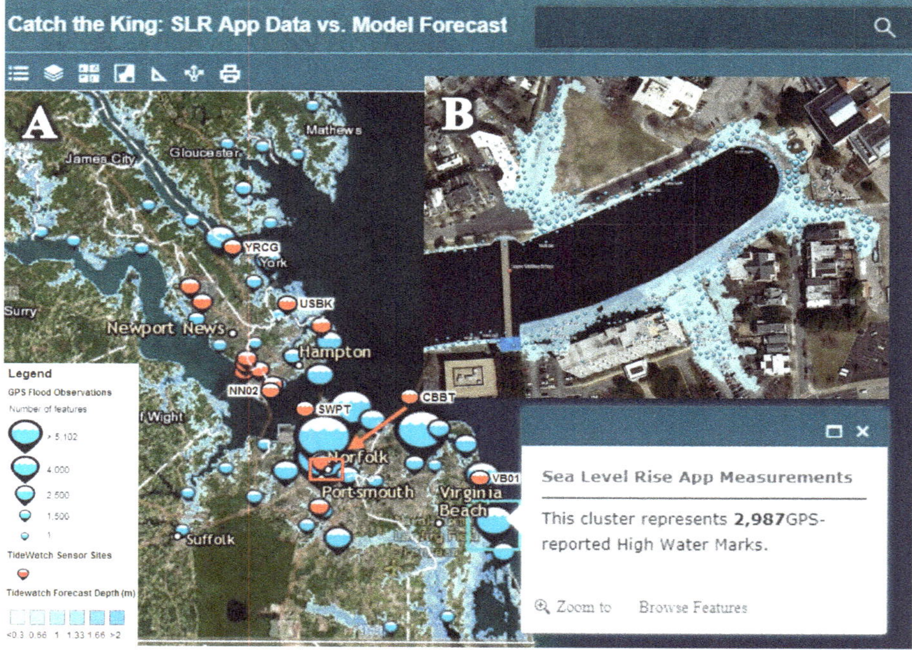

Figure 6. (**A**) Overview of study region in Hampton Roads, VA, featuring the 2017 king tide maximum inundation forecast from the Tidewatch Map in blue, GPS citizen science observations as blue dots, and water level sensors from the Tidewatch Charts as red dots. (**B**) Inset shows a high-density concentration of flood validation data in the historic Hague region of Norfolk, VA, USA, where the model had favorable agreement with the citizen science observations. Here, the model yielded a 94% spatial match, and a 4.2 m MHDD. See the map online: http://arcg.is/1HLOPS.

3. Results

Spatial data collected for each king tide event were aggregated through the Sea Level Rise mobile app and shared online using interactive web maps, so that volunteers with minimal GIS experience could visualize their own GPS observations populate the Tidewatch model's predictions in near-real time (Figure 6). This level of data accountability implemented an open interaction where users could conduct their own analyses and make their own mean difference calculations through ArcGIS Online's distance and area measuring tools from their data in a web browser while viewing the public map. This data interactivity spurred high engagement and participation for students involved in STEM research or related educational classes. Figure 6A shows an aggregated point map of 59,718 high water marks superposed with the Tidewatch Maps throughout the greater Hampton Roads region and showcases the extent of areas not covered by automated sensors that were surveyed through *Catch the King* in 2017 (Table 1).

Comparatively, 2018 surveyed an even broader area than was monitored in 2017, but with less density (Figure 4). For example, Figure 6B shows Norfolk's Hague, where thousands of GPS data were collected in both years. In areas where a significant point density is reported, data points can form more than simple flood contours when combined with digital elevation survey data. With sufficient point density, one can form their own observation-driven interpolated flood model for comparison with hydrodynamic simulation results. In this case, the areal extent of The Hague shown in Figure 6B yielded a 94% match with the raster polygon built from the interpolated GPS points, with the Tidewatch model slightly erring on the side of over-prediction. The cursory overview of the Greater Hampton Roads Region shown in Figure 6 shows the Tidewatch Map in blue, GPS citizen science observations as aggregated blue dots, which render and dis-aggregate based upon the zoom level, and water level sensors from the Tidewatch Charts as red dots. This legend theme will persist through the next several spatial comparison figures, and was designated through the Sea Level Rise app for the observation data, and for the model via a meeting of the CCRFR with emergency managers [44].

Table 1. A tabulated table of quantities mapped by *Catch the King* volunteers in coastal Virginia, USA, where the SCHISM-Driven Tidewatch storm tide model provided inundation forecasts.

Year	Distance (Unique km)	Distance (Total km)	GPS Points	Photographs
2017	104.28	416.66	59,718	1582
2018	69.36	277.77	37,728	458
TOTAL	173.64	694.43	97,446	2040

Given that flood impacts for king tides are simply tidal calibration data that are likely to be aligned along similar elevation contours without intervening atmospheric conditions, linear distance metrics can be useful to compute spatial differences in relatively flat areas. The standard distance formula may then be computed by GIS software to calculate the difference between each GPS data point to the nearest predicted inundated space. To compute this, the modeled geospatial flood depths served through VIMS' Tidewatch Maps were converted into vector data polygons, with the maximum flood extent representing the 0 m flood contour. As the volunteers were instructed to map inundation in their communities by dropping time-stamped digital GPS breadcrumbs, the citizen scientists' data should ideally represent the observed GPS flood extents, and in most places, the model had an overwhelmingly favorable agreement. Figure 7 shows an example in Norfolk's Larchmont neighborhood adjacent to a dog park, where flooding is frequent, and the 112 points were used to compare with the light blue modeled flood extents as a linear distance, and averaged to form a mean horizontal distance deviation (MHDD) metric, which yielded an average deviation of 2.67 m for this site during the 2018 king tide.

Likewise, several other areas, ranging from residential, commercial, and industrial land uses, during the 2017 king tide are featured in Figure 8. Since Tidewatch Maps provide more than simply a maximum inundation extent, unlike tidal depths estimated from a bathtub model or a sea level rise topographic flood elevation viewer, temporal accuracy can also be assessed through the GPS timestamps reported on each user's measurements through the Sea Level Rise app. Figure 8A,C,E,G, shows a model distance comparison of forecasts and data from 13:00 to 13:59 UTC. Figure 8B,D,F,H show observation data and model forecasted depths for the same sites an hour later from 14:00 to 14:59 UTC. These figures are used to show varying flooding conditions during the king tide on 5 November 2017, which occurred at 14:30 UTC, similar to those noted in Figure 3 during 2011 Hurricane Irene. Figure 8 shows individual sites where monitoring efforts took place in 2017 and ultimately contribute to the overall figure of +/−5.9 m in horizontal difference between the maximum extents predicted via the Tidewatch Maps and the 59,718 high water marks measured through *Catch the King*.

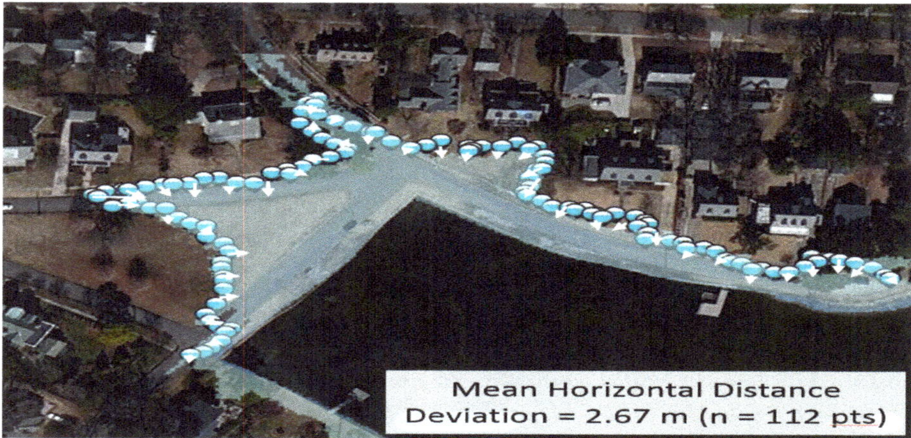

Figure 7. Illustrative comparison of mean distance difference of 112 GPS data points collected in Larchmont, Norfolk (2.67 m), compared with the 2018 king tide peak inundation forecast in blue.

Figure 8A shows data from the 2017 king tide for the same area as Figure 7 in Larchmont, Norfolk for 44 high water marks collected by 2 citizen scientists from 13:32 to 13:52 UTC to yield a MHDD = 4.18 m. The same site from an hour later is shown in Figure 8B, comprised of 61 high water marks collected by 1 volunteer from 14:30 to 14:39 UTC to yield a less favorable MHDD = 6.92 m during the peak inundation period in *Catch the King* 2017. Figure 8C depicts inundation during the 2017 king tide along the south bank of the Lafayette River near the Haven Creek Boat Ramp in Norfolk by 73 high water marks collected by 2 citizen scientists from 13:35 to 13:59 UTC to yield an MHDD = 4.67 m. The same site an hour later is shown in Figure 8D, this time featuring 136 high water marks collected by 3 people from 14:30 to 14:47 UTC to yield a less satisfactory MHDD = 6.29 m during the peak inundation period on 5 November 2017. Figure 8E showcases GPS data from *Catch the King* 2017 for the Lafayette Shores neighborhood, nestled in the east bank of the Lafayette River, through six high water marks collected by a citizen scientist from 13:49 to 13:52 UTC to yield an MHDD = 2.15 m. The same site from an hour later is shown in Figure 8F, comprised of 29 high water marks collected by 2 people from 14:30 to 14:47 UTC to yield a less favorable MHDD = 4.70 m during the peak inundation period. Figure 8G depicts inundation during the 2017 king tide along the north bank of Little Creek in the East Ocean View neighborhood of Norfolk via 93 high water marks collected by 2 citizen scientists from 13:44 to 13:56 UTC to yield an MHDD = 9.81 m. The same site from one hour later is highlighted in Figure 8H, now featuring 68 high water marks collected by 1 person from 14:40 to 14:45 UTC to yield an improved MHDD = 4.06 m during the peak inundation period during *Catch the King* 2017.

While the areas shown in Figure 8 were surveyed by few citizen scientists, the area shown in Figure 6B is one of the most frequently monitored areas in the Sea Level Rise app's history. During 2017's *Catch the King*, the area featured in Figure 6B was monitored by 27 different volunteers at different times (not all during the flood peak period) to form 27 king tide inundation contours for The Hague. These were mosaicked into a composite maximum extent contour map comprised of 1134 GPS points stretching 2.17 km to form a maximum extent contour for VIMS to compare with its Tidewatch Maps modeled inundation. The total distance walked and recorded using the Sea Level Rise app by all 27 volunteers for The Hague alone in 2017's *Catch the King* was 22.58 km (Figure 6B). This is 10.39× the composite's distance at this king tide inundation site, meaning >10× the actual effort, or about a 10× greater distance was walked than represented by the composite 0 m flood depth contour. As a result, these distances along the waterways that were travelled as effort expended by volunteers was significantly greater than needed to efficiently validate the flooding extents (by 10×), and this is not counting a volunteer's travel to and from each reported flood site.

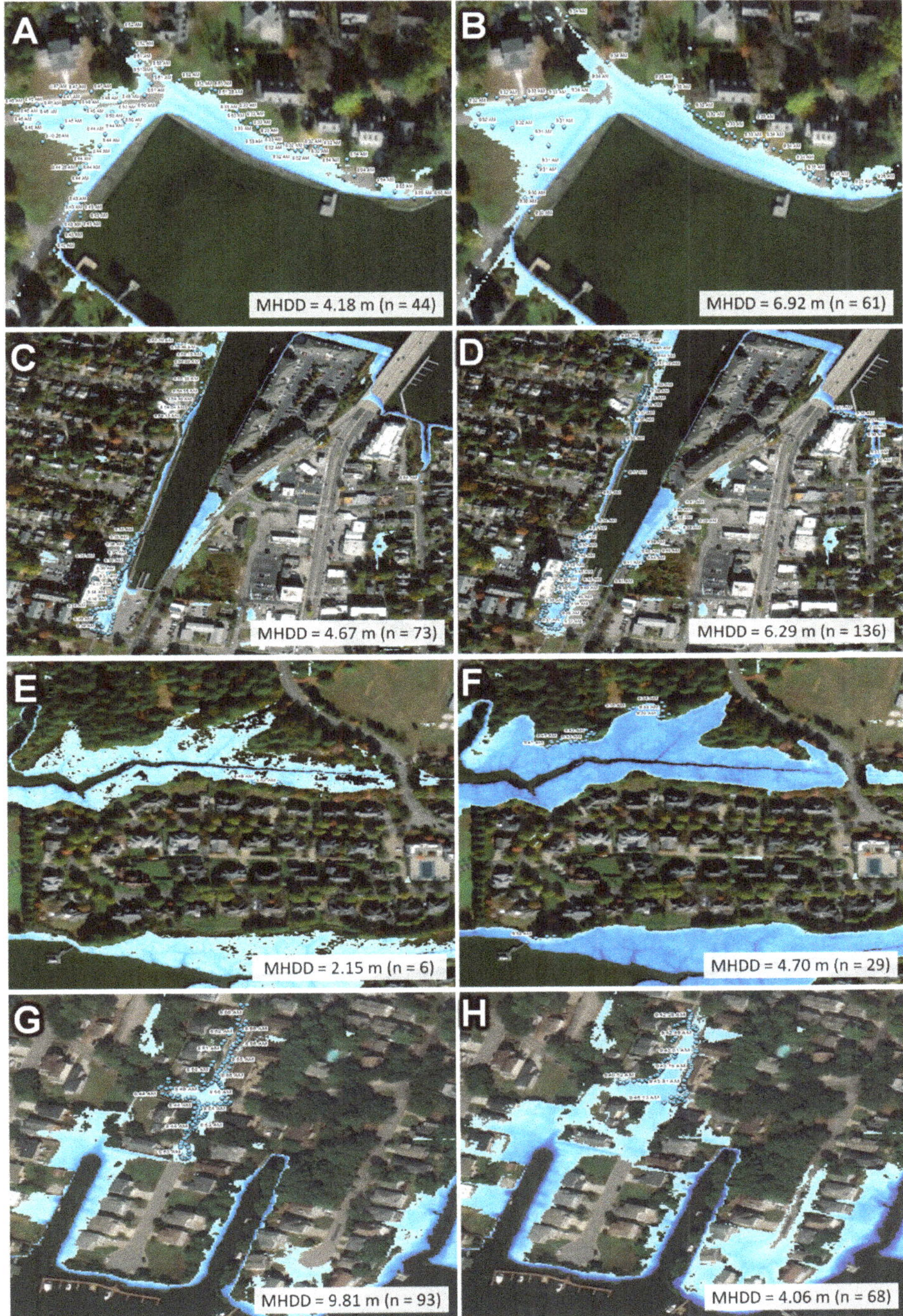

Figure 8. Model comparison of forecasts and data from 13:00 to 13:59 UTC at left, and 14:00 to 14:59 UTC used to show varying flooding conditions during the 2017 king tide at four sites ranging from residential, commercial, and industrial land uses throughout Norfolk, VA.

For perspective, the grand total for over 1000 volunteers to map inundation distances across both *Catch the King* years was 631.35 km, with the total number of unique king tide contours travelled in Hampton Roads across both years adding up to 173.65 km (Table 1). Therefore, significantly more effort was used than needed to effectively map the site, with Norfolk's Hague experiencing the greatest duplicated effort. This was also indicative of overlapping efforts among other high-density monitoring areas at public beaches in Norfolk, Virginia Beach, and Hampton representing the other greatest density GPS data density areas with >6× overlap. As noted previously, the increased density of GPS data proved to be a boon towards supporting the development of data-driven area maps, which was useful in The Hague, but were less useful on public coastal beaches where the water was not surrounded by land, with transient sand elevations that may vary from those embedded in the hydrodynamic model via the latest lidar elevation surveys. Thus, in 2018's *Catch the King*, greater emphasis was placed on coordinating volunteers at registration to commit to mapping unique locations when communicating with volunteers via training events, and through print and social media to best value their time commitment and most efficiently validate the model.

Aside from the horizontal GPS surveys reported through *Catch the King*, Tidewatch is routinely validated through automated water level monitoring sites. An overview of the water level sensor data extracted from sensors through Tidewatch Charts during *Catch the King* 2018 across all data points revealed a favorable average vertical accuracy assessment of 3.7 cm via the root mean squared error (RMSE). This metric was drawn from 28 StormSense water level sensors, and 16 tidal USGS Sensors and 4 NOAA sensors. Six of these sensors, including three NOAA, two StormSense, and one USGS sensor, are shown in Figure 9 from VIMS' Tidewatch Charts, as an example comparison of hydrodynamic model performance during 2018's *Catch the King*. These charts from 2018 are labeled in Figure 6A with their four-character station abbreviations, for spatial reference, with their time series data shown in Figure 9.

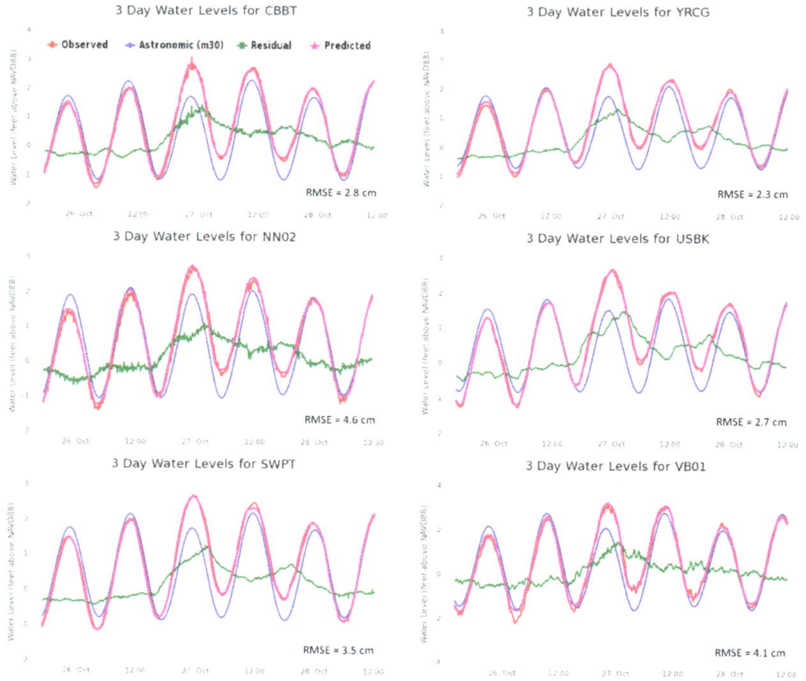

Figure 9. Time series of six Tidewatch Charts' water level sensors' observations (red), astronomical tidal estimations (blue), and computed residual difference (green), compared with SCHISM model predictions (pink) from Tidewatch Maps during *Catch the King* at 10:30 on 27 October 2018; UTC-5.

The region had 16 less water level sensors in 2017, and *Catch the King* in 2017 took place during a king tide with no additional amplifying wind or rainfall effects. The aggregate RMSE comparison in 2017 across 32 sensors was 3.5 cm, resulting in a slightly better agreement with the model than the 3.7 cm RMSE value reported in 2018 [27,35]. As a result of the "blue sky" conditions, 722 citizen scientists collected data in 2017, but their data was less dynamically interesting than 2018, which had less volunteers (431), due in part to a mild nor'easter that occurred on the night before *Catch the King*, making the weather less favorable for volunteers. The nor'easter brought 11.17 m/s (25 mph) sustained winds for nearly 3 h from 03:00 to 06:00 UTC on 27 October 2018 (yet contributed negligible rainfall), as seen in the residual fluctuations represented by the green line of each automated monitoring gauge's measurements in Figure 9.

4. Discussion

As citizen scientists contribute significant amounts of their time to collect these intricate geospatial data sets, care is taken by the custodians of those data to perform quality assurance and quality control on those data before subjecting them to rigorous scientific analysis. The raw volunteer data for each *Catch the King* survey in 2017 and 2018 were modified after initial statistics were reported to filter out and otherwise minimize bias in this study. This was in an effort to provide the most meaningful model validation statistics, which were reported in the results section, and were honed to validate three important factors for model validation: Duration, depth, and degree of inundation:

Duration

(1) Points with a reported timestamp more than an hour outside of the time window in which the king tide occurred at that location were not included in the comparison.

(2) Those points within the window were rounded to the nearest hour to split them into comparative groups for each hourly model output for comparison (as depicted in Figure 8).

Depth

(3) Surveyed points were merged with the topobathymetric DEM used to build the model, developed by the USGS and published in [45], to append elevation values.

(4) Any points >0.91 m (3 ft) elevation above the North American Vertical Datum of 1988 (NAVD88) were not included in the model comparison, as the king tide from neither year exceeded this height at any water level sensor in the region (Figure 7).

Degree

(5) Points with an appended photograph were collected back away from the water's edge (for land marking and visual perspective), and thus were not included as part of the flood contour comparison. If included, these would over-predict the degree of flooding.

(6) Points with a radial accuracy metric >10 m (32 ft) were too inaccurate to include, as they misrepresent the degree an area was inundated, favoring over-prediction (Figure 10).

The extent that duration of inundation was addressed and timing of when a flood event will arrive dictates the potential mitigating actions that may be taken. Tidal inundation events can easily be predicted through harmonic algorithms, and hydrodynamic models can improve upon this by informing citizen scientists, community planners, and emergency managers alike when the flood waters will arrive. This information is useful for personal preparation of one's home and assets that may be in low lying areas, route planning and guidance for personal and emergency response vehicles, and for scheduling road closures to minimize vehicular loss. Figures 4 and 8 illustrate the difference an hour makes in terms of accuracy on model validation, and in the future, the recommendation for more frequent spatial mapping has already been recommended for the future development of Tidewatch Maps to eventually shift to 30 min time steps for the online time-aware layers for depicting more temporally-resolute flood mapping beyond hourly updates. However, presently, Tidewatch Maps are used to map all of Virginia's coastal floodplain via SCHISM's model outputs at a 1 to 5 m resolution (depending upon the accuracy of lidar point spacing for the model's underlying digital elevation assumptions). A 36-h Tidewatch forecast already consists of 37 state-wide coastal flood maps being automatically produced every 12 h. Thus, doubling that number to 72 iterative Tidewatch Maps per cycle is both computationally expensive for the model's post-processing, and impractical for users loading its flood forecasts via the web without newer technology to enhance loading times. Since users have most frequently accessed Tidewatch Maps using their smart phones to view its flood predictions during periods of significant power and internet outages, greater temporal resolution for 30-min update intervals is not likely to be implemented soon, as additional loading times are even more cumbersome for mobile devices.

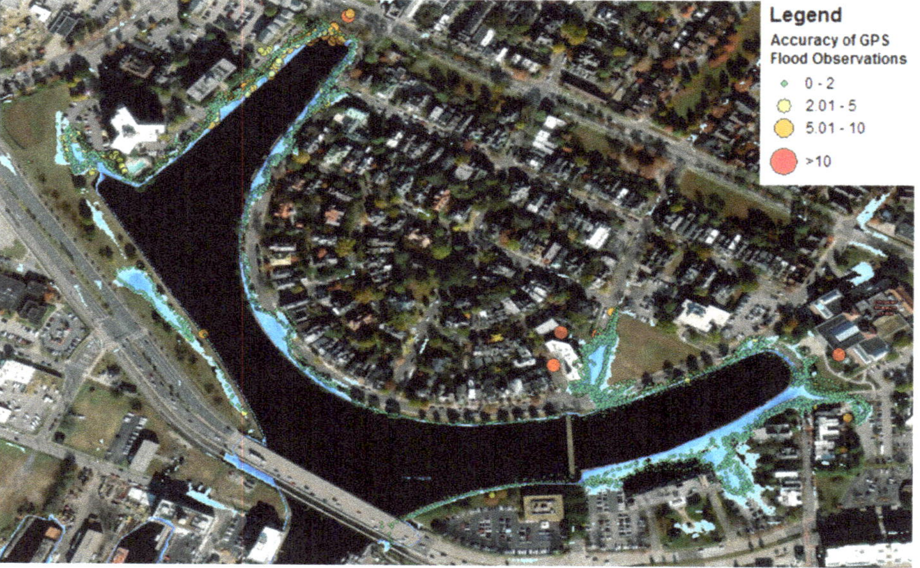

Figure 10. Map of radial positional accuracy (in meters) reported by the Sea Level Rise mobile app in Norfolk's Hague during *Catch the King* 2017. Points in red were not included in the spatial comparison.

Aside from depths being directly validated via amplitude comparisons with automated water level sensors, surveyed points collected through *Catch the King* were merged with a DEM to translate the collected data to contours. While most modern smart phones have an altitude sensor, its error on accuracy is not sufficient to accurately report flood depths or meaningfully report heights above a reference datum. The Sea Level Rise app does display one's altitude in the app interface and records this with each point, but not all phone models share these data with the app or possess the internal hardware to report this. Thus, for the most reliable elevations, the GPS high water marks were merged with the topobathymetric DEM used to build the SCHISM model and the Tidewatch Maps, developed by the USGS [45]. Many citizen scientists were likely to test the app before venturing out to collect data, and several data points that were nowhere near water were collected. Instead, these points appeared in houses, apartment complexes, or traced around isolated puddles in parking lots that were non-contiguous with neighboring estuaries. These locations were flagged for use in storm water studies, and removed from this tidal inundation model validation analysis for any points >0.91 m (3 ft) elevation above NAVD88 were not included in the model comparison, as the king tide from neither year exceeded this height at any water level sensor in the region, and there was no significant rainfall

(>2 cm) accumulated during or preceding either *Catch the King* tidal flood mapping event. In other cases, users mapped tidal-connected drainage ditches that became inundated during the king tides and these points were included in the analysis (Figures 11 and 12).

The context through which the degree of inundation was monitored by citizen science data is made more useful through following proper training for data collection and appropriate data filtering. Data were collected by 722 volunteers in 2017 and 431 volunteers in 2018. These data were collected by over 20 different smart phone models, which each vary in terms of relative accuracy due to the number of antennae included in each phone model to aid in triangulation of positional accuracy and for general clarity of cellular broadband communications through the device. As such, citizen science surveys are inherently less precise than those conducted by professional scientists using industry-standard GPS receivers capable of real time kinematics (RTKs). Since the high variation in phone models introduces variable accuracy, as does the number of GPS satellites in range, an estimated radial error metric is reported by the Sea Level Rise app for each GPS measurement by assessing the incoming signals from the global navigation satellite systems along with a correction stream. However, unlike professional survey equipment operated by trained professionals, smart phones are presently unable to achieve the 1 cm positional accuracy that RTK GPS tools can. Thus, points with a radial accuracy metric >10 m were not included in the spatial comparison (Figure 10).

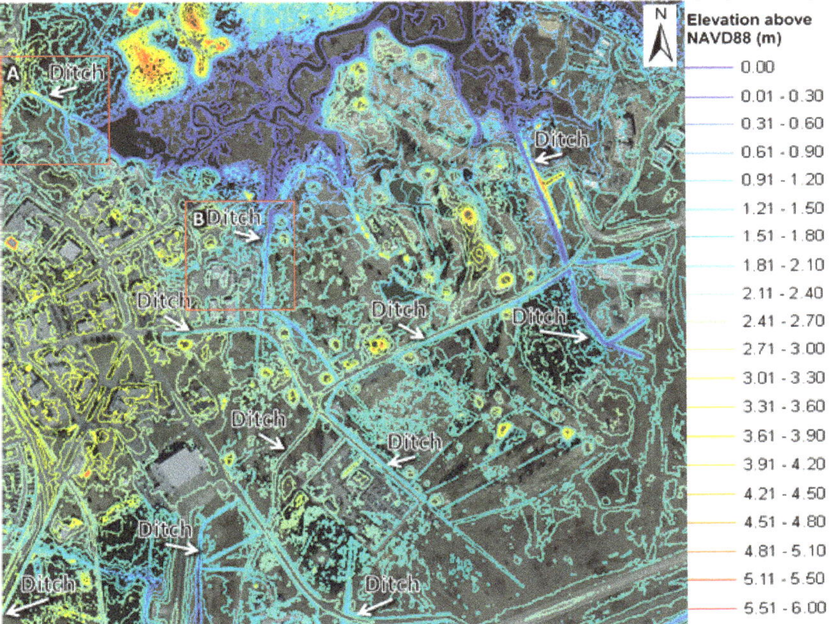

Figure 11. Examples of obtaining aerial elevations from lidar and conducting hydro-correction to assure unobstructed ditch and creek channels persist for hydrologic transport of king tide inundation in the City of Hampton, VA. Red highlighted boxes correspond to areas depicted in Figure 12A,B.

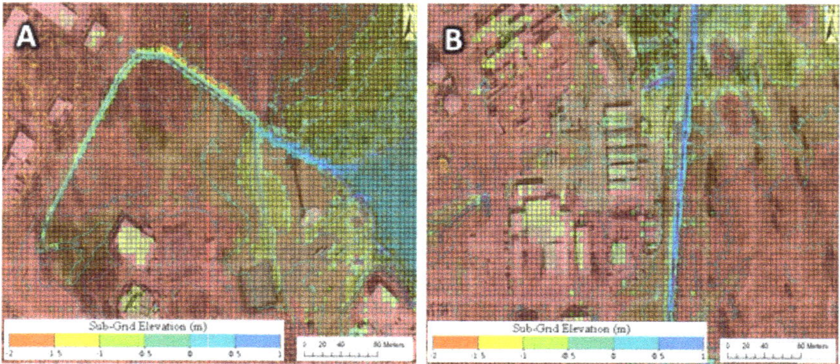

Figure 12. Ditches extracted and represented at the sub-grid DEM pixel level for effective representation of drainage ditches leading to (**A**) the west edge of Tabbs Creek and (**B**) the south edge of the creek draining the fairways of Eaglewood Golf Course, both identified for hydro-correction in Hampton, VA, USA.

Upon filtering for these three things, it was found that the Tidewatch Map comparisons on 5 November 2017 during *Catch the King* 2017 had an overall MHDD of 5.9 m (19.3 ft). This statistic was calculated from 57,986 of the 59,718 total high water marks collected after less than 3% of the citizen scientists' measurements were filtered out for any of the six reasons previously noted for relative error on duration, depth, or degree of flooding. In a similar fashion, comparisons between the high water marks collected by citizen scientists during *Catch the King* 2018 observed a slightly less favorable overall MHDD of 6.2 m (24.6 ft), likely attributed to the winds from the mild nor'easter that occurred in the hours leading up to the event. This MHDD was calculated from 30,920 of the 33,847 total high water marks collected after 8.6% of the citizen scientists' measurements were filtered out of the surveyed data.

In the interest of improving future forecasts, it was found that less than 1% of the filtered GPS high water marks were still not within 50 m of the Tidewatch Map's predicted inundation raster. Further investigation into these sites identified two reasons for the discrepancy, both related to errors in hydrologic correction of the model's DEM calculated water depth assumptions. Figure 11 outlines a series of above-ground drainage ditches in Hampton VA, that occasionally become inundated when the water table rises with extra high tidal waters. Connection through these narrow drainage ditches can be obscured by thick canopied trees adjacent to the narrow tidal creeks and mostly non-tidal ditches that feed those creeks (Figure 12). The model's elevations are attributed to averaged digital elevations from aerial lidar surveys to source the DEM that the model uses to represent reality. Thus, the depths of the bottoms of these fine scale ditches (<1 m wide) were not likely to be correct unless the point spacing is extremely high. Naturally, this is acceptable, since the model was scaled to (at best) 1 m spatial resolution, and cannot accurately represent the slopes of such detailed drainage features without scaling to a 0.33 m resolution. Yet, these ditches were found to become tidal conduits for fluid movement capable of causing inundation far from the shoreline during king tides [46]. In other places, bridges over typically non-tidal creeks were not removed from the aerial survey data used to build the DEM, and removal of the occluding feature aided hydro-correction to correct the model's incorrect volume displacement in areas where entire creeks were shown to be dry due to the artificial dam imposed by a bridge, constricting proper fluid flow (Figure 13). Thus, one of the most important and immediately noticeable achievements that *Catch the King* accomplished for the hydrodynamic model's validation was the aid of hydro-correction for several small streams that were obscured in the aerial lidar surveys informing the Tidewatch Maps. In the case of several ephemeral creeks that temporarily became tidal during the king tide, the citizen scientists' survey identified locations where these ditches needed to be corrected (Figure 14) [47].

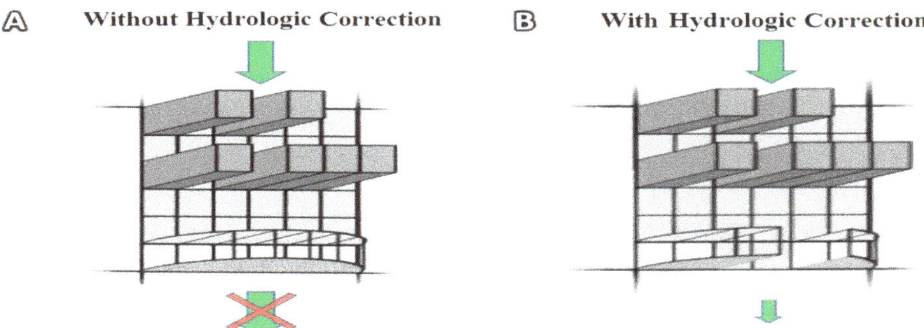

Figure 13. Conceptualization of macro-roughness features in an urban environment resolved within the street level inundation forecasts. In the example shown in (**A**), a bridge artificially obstructs flow from passing through to the hydrodynamic model grid cell below the one shown. With hydrologic correction provided by the citizen scientists via *Catch the King* (**B**) shows the grid opened, where a nearby flood contour can be extracted and applied to translate the inundation underneath the bridge and open flow to the opposite side, no longer impeding fluid flow.

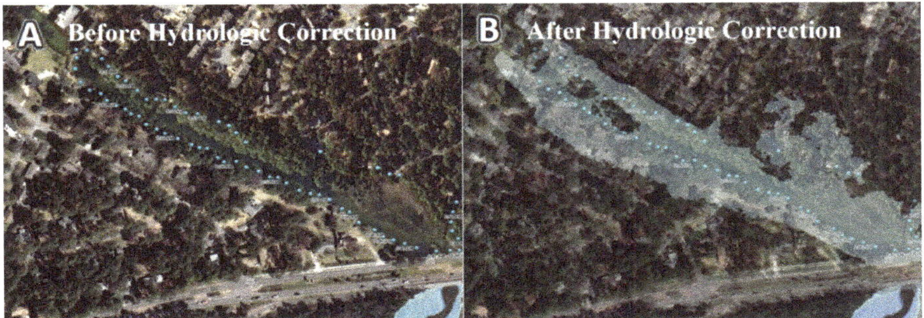

Figure 14. Citizen Science flood extent observations aided in hydrologic correction for an ephemeral stream feeding Wolfsnare Creek in Virginia Beach, VA. (**A**) Citizen scientists mapped the tidal inundation extent approximately an hour before the king tide's peak. (**B**) Hydrologic correction fixed the lidar-derived DEM to permit flow through the small box culvert beneath the bridge to enhance the model's spatial accuracy via better estimation of cross-sectional flow and volume conservation.

For example, a typically non-tidal creek feeding Wolfsnare Creek in Virginia Beach was inundated during the king tide in 2017. *Catch the King* volunteers mapped the king tide approximately an hour before the king tide's peak, and the large initial mean horizontal distance difference from this cluster of points drew researchers' attention to investigate the hydrodynamic model's under-prediction of inundation. The error was traced back to a faulty elevation assumption attributed to obstructed flow underneath a bridge. VIMS researchers hydro-corrected the landscape to open flow using neighboring elevations from the DEM through the box culvert underneath the bridge and corrected ground elevations impacted by thick tree canopies surrounding a creek bed with low aerial lidar-point-spacing.

5. Conclusions

A large-scale flood monitoring citizen science data collection effort was used to favorably validate an automated browser-based flood mapping service driven by a cross-scale hydrodynamic model predicting storm tide inundation in coastal Virginia, USA. The operational modeling effort for predicting tidal flooding can be mapped using multiple methods, yet the most effective method was found to be the automated implementation of a street-level hydrodynamic model. The Tidewatch Maps implemented by the Virginia Institute of Marine Science (VIMS) leveraged their SCHISM hydrodynamic model with inputs of: atmospheric wind and pressure data, tidal harmonic predictions at the open boundary, and prevailing ocean current inputs, such as the Gulf Stream. This information was successfully computed from a large scale model and translated to the street-level via SCHISM's computationally efficient non-linear solvers, and semi-implicit numerical formulations aided by a sub-grid geometric mesh with embedded lidar elevations.

Validation in the vertical scale found that the SCHISM model outputs via the Tidewatch web mapping platform compared well in Hampton Roads among the 32 extant water level sensors during the highest astronomical tide of the year on 5 November 2017, a king tide, yielding an aggregate RMSE of 3.5 cm. The region expanded its sensor base to 48 through an IoT sensor project, StormSense, to compare well again during the king tide on 27 October 2018, resulting in an RMSE of 3.7 cm. Horizontal validation was aided by time-stamped GPS flood extent data collected by citizen scientists through the world's largest environmental survey (in terms of the most contributions in the least amount of time), *Catch the King*. The citizen science flood mapping survey was established in Hampton Roads in 2017 and recruited volunteers through local, print, and social media outlets. The survey's organizers then trained the citizen scientists in the use of the free Sea Level Rise mobile flood mapping application at frequently inundated public spaces in the months leading up to each king tide event.

The citizen scientists' flood monitoring data formed time-indexed GPS breadcrumbs to form contours that were successfully aggregated and compared with the maximum inundation extents of the same time interval from VIMS' Tidewatch Maps. The data were filtered to minimize bias attributed to errors related to observing flooding duration, depth, and degree. Once the *Catch the King* survey data were filtered for these three things, it was found that the Tidewatch Map comparisons on 5 November 2017 had an overall mean horizontal distance difference of 5.9 m (19.3 ft). The model comparison with the observations collected during the king tide on 27 October 2018 were found to be less favorable, yielding an average distance deviation of 6.2 m (24.6 ft), likely attributed to the winds from the mild nor'easter that occurred in the hours leading up to the event. In each spatial validation effort, less than 9% of the surveyed data were excluded from the analysis.

Lessons learned from citizen science surveys have improved the model through cost-effective hydrologic correction of mission conduits for fluid flow. These were identified by filtered GPS observations that the model missed in its initial automated forecast, but were corrected in hindcast, in preparation for the next significant inundation event. Errors in hydro-correction did not relate to errors in friction parameterization of the model, but were more associated with flow pathways that were occluded from aerial lidar surveys. These areas included bridges, culverts, and stormwater drainage systems without tidal backflow prevention valves, which formed artificial dams in the digital surface model embedded in the forecasted Tidewatch Maps. Many of these identified areas have been corrected and have recently been used alongside the successful model validation in Hampton Roads to expand the forecast area of the Tidewatch Maps beyond southeast Virginia to include the entire coastal zone of Virginia in 2019.

As king tides are currently simply nuisance floods, which primarily inundate streets and driveways without significantly damaging infrastructural assets, the issues are presently geared towards traffic and transportation issues. Common concerns from citizen scientists involved in the *Catch the King* mapping events involved concerns regarding whether their vehicle could be safely street parked or if their vehicles needed to be safely moved into a garage during king tides. Others questioned whether they should take an alternate route to work or school or the store due to potential street flooding. As technology progresses, these questions will become more prevalent as we aim to ascertain whether modern route guidance mobile applications will be intelligent enough to account for intermittent inundation, or unintentionally lead vehicles down flooded streets simply because there is no traffic detected on them while an adjacent elevated street is congested. Some navigation applications, such as Waze, have aimed to crowdsource all road hazard data through their "Connected Citizens" program, but this method is only a temporary solution, as a model cannot currently automate road hazard flags for flooded locations where those particular app users are not or have not logged data.

Naturally, adeptly answering these questions becomes increasingly difficult once self-driving vehicles are involved. Thus, the outcomes of inundation modeling efforts for this tidal calibration effort will more significantly be realized once this trained citizen scientist army is deputized into post-hurricane surveys. Since 2011 Hurricane Irene was the last hurricane to significantly impact Virginia's Hampton Roads region, the Tidewatch automated mapping model has yet to demonstrate widespread accuracy amidst a significant inundation event since the Sea Level Rise app's advent in 2014. The goal is to continue to improve the model with each *Catch the King* tidal calibration and train volunteers so they will be aware of where to find the latest flood forecast information, and how to collect meaningful flood validation data. Thus, this monitoring coordination approach with hydrodynamic modeling provided a novel procedural release of information to depict predicted maximum inundation extents for expediently effective model validation through the use of an overwhelming quantity of quality event data with relatively low risk to volunteer citizen scientists.

Author Contributions: Conceptualization, J.D.L., H.V.W., D.M., and W.A.S.; experiments conduction, J.D.L. and D.R.F.; formal analysis, J.D.L., D.S.; writing—original draft preparation, J.D.L. and M.M.; writing—review and editing, J.D.L.

Funding: VIMS' flood modeling research for this study was funded by the CCRFR. In 2018, *Catch the King* was financially sponsored by WHRO Public Media, the Hampton Roads Sanitation District, and school group involvement was encouraged through a grant from the Hampton Roads Community Foundation, and the Batten Environmental Education Initiative. In 2017, *Catch the King* was funded by The Virginian-Pilot, The Daily Press, WHRO Public Media, and WVEC-TV. Funding for this article's publication was graciously furnished by the CCRFR and through FEMA grant #FEMA-DR-4291-VA-011 to VIMS through the Virginia Department of Emergency Management.

Acknowledgments: The authors would like to thank the three reviewers whose kind suggestions and recommendations greatly improved the quality of the final published manuscript. *Catch the King* is supported by WHRO Public Media, The Virginian Pilot, the Daily Press, WVEC News 13, and the CCRFR. *Catch the King* also is made possible by the nonprofit groups, Wetlands Watch and Concursive Corp., creators and developers of the citizen-science 'Sea Level Rise' mobile app (freely available on iOS and Android platforms). *Catch the King's* adept volunteer coordinator is Q.J. Without her organization and the significant community support of volunteers, this effort would not nearly have reached the success that it has realized; a list of their names is available online: https://bit.ly/2ZLZwBS.

Conflicts of Interest: The authors declare no conflict of interest.

References

1. Johnson, B.H.; Kim, K.W.; Heath, R.E.; Hsieh, B.B.; Butler, H.L. Validation of three-dimensional hydrodynamic model of Chesapeake Bay. *J. Hydraul. Eng.* **1993**, *119*, 2–20. [CrossRef]
2. Blumberg, A.F.; Khan, L.A.; St. John, J.P. Three-dimensional hydrodynamic model of New York Harbor region. *J. Hydraul. Eng.* **1999**, *125*, 799–816. [CrossRef]
3. Lesser, G.R.; Roelvink, J.V.; Van Kester, J.A.T.M.; Stelling, G.S. Development and validation of a three-dimensional morphological model. *Coast. Eng.* **2004**, *51*, 883–915. [CrossRef]
4. Loftis, J.D.; Mitchell, M.; Atkinson, L.; Hamlington, B.; Allen, T.R.; Forrest, D.; Updyke, T.; Tahvildari, N.; Bekaert, D.; Bushnell, M. Integrated Ocean, Earth and Atmospheric Observations in Hampton Roads, Virginia. *Mar. Technol. Soc. J.* **2018**, *52*, 68–83. [CrossRef]
5. Dorman, J.K.; Banerjee, N. A Real-Time Flood Warning System. *ArcUser* **2016**, *19*, 16–19.
6. Loftis, J.D.; Katragadda, K.; Rhee, S.; Nguyen, C. StormSense: A Blueprint for Coastal Flood Forecast Information & Automated Alert Messaging Systems. In Proceedings of the 3rd International Workshop on Science of Smart City Operations and Platforms Engineering, SCOPE'18, Porto, Portugal, 10 April 2018; p. 3.
7. Loftis, J.D.; Forrest, D.; Katragadda, K.; Spencer, K.; Organski, T.; Nguyen, C.; Rhee, S. StormSense: A New Integrated Network of IoT Water Level Sensors in the Smart Cities of Hampton Roads, VA. *Mar. Technol. Soc. J.* **2018**, *52*, 56–67. [CrossRef] [PubMed]

8. Loftis, J.D.; Wang, H.; Forrest, D.; Rhee, S.; Nguyen, C. Emerging Flood Model Validation Frameworks for Street-Level Inundation Modeling with StormSense. In Proceedings of the 2nd International Workshop on Science of Smart City Operations and Platforms Engineering, SCOPE'17, Pittsburg, PA, USA, 21 April 2017; pp. 13–18.

9. Loftis, J.D.; Taylor, G. Leveraging Web 3D for Street-Level Flood Forecasts. *ArcUser* **2018**, *21*, 22–25.

10. Boon, J.D.; Brubaker, J.M. Acoustic-microwave water level sensor comparisons in an estuarine environment. In Proceedings of the OCEANS 2008, Quebec City, QC, Canada, 15–18 September 2008; pp. 1–5.

11. Simoniello, C.; Jencks, J.; Lauro, F.; Loftis, J.D.; Weslawski, J.M.; Deja, K.; Forrest, D.R.; Gossett, S.; Jeffries, T.; Jensen, R.M.; et al. Citizen-Science for the Future: Advisory Case Studies from Around the Globe. *Front. Mar. Sci.* **2019**. [CrossRef]

12. Loftis, J.D.; Mayfield, D.; Forrest, D.; Stiles, W. A Geospatial Analysis of +50,000 Citizen-Science collected GPS Flood Extents and Street-Level Hydrodynamic Model Forecasts during the 2017 King Tide in Hampton Roads, VA. In Proceedings of the MTS/IEEE Oceans 2018, Charleston, SC, USA, 22–25 October 2018.

13. Blumberg, A.F.; Georgas, N.; Yin, L.; Herrington, T.O.; Orton, P.M. Street-scale modeling of storm surge inundation along the New Jersey Hudson river waterfront. *J. Atmos. Ocean. Technol.* **2015**, *32*, 1486–1497. [CrossRef]

14. Wang, H.; Loftis, J.D.; Forrest, D.; Smith, W.; Stamey, B. Modeling Storm Surge and inundation in Washington, D.C.; during Hurricane Isabel and the 1936 Potomac River Great Flood. *J. Mar. Sci. Eng.* **2015**, *3*, 607–629. [CrossRef]

15. McCallum, B.E.; Wicklein, S.M.; Reiser, R.G.; Busciolano, R.; Morrison, J.; Verdi, R.J.; Painter, J.A.; Frantz, E.R.; Gotvald, A.J. *Monitoring Storm Tide and Flooding from Hurricane Sandy Along the Atlantic Coast of the United States, October 2012*; U.S. Geological Survey Open-File Report 2013–1043; Office of Surface Water, U.S. Geological Survey: Reston, VA, USA, 2013; p. 42.

16. Loftis, J.D. Catch the King Tide Thank You and Review. In Proceedings of the CCRFR Thank You and Review Community Event at ODU, Norfolk, VA, USA, 13 December 2017; Presentations 41. [CrossRef]

17. Rumson, A.G.; Hallett, S.H. Innovations in the use of data facilitating insurance as a resilience mechanism for coastal flood risk. *Sci. Total Environ.* **2019**, *661*, 598–612. [CrossRef] [PubMed]

18. Rogers, L.; Borges, D.; Murray, J.; Molthan, A.; Bell, J.; Allen, T.; Bekaert, D.; Loftis, J.D.; Wang, H.; Cohen, S.; et al. NASA's Mid-Atlantic Communities and Areas at Intensive Risk Demonstration: Translating Compounding Hazards to Societal Risk. In Proceedings of the Oceans 2018, Charleston, SC, USA, 22–25 October 2018.

19. Wang, R.Q. Big Data of Urban Flooding: Dance with Social Media, Citizen Science, and Artificial Intelligence. *EGU Gen. Assem. Conf.* **2018**, *20*, 404.

20. VA Catch the King Tide. 2019. Available online: https://www.vims.edu/kingtide/ (accessed on 6 June 2019).

21. NC King Tides Project. 2019. Available online: http://nckingtides.web.unc.edu/ (accessed on 6 June 2019).

22. Witness King Tides Project. Australia National King Tide Monitoring Program. 2019. Available online: http://www.witnesskingtides.org/ (accessed on 6 June 2019).

23. NOAA King Tides Project. National Oceanic and Atmospheric Administration's Office for Coastal Management. 2019. Available online: https://arcg.is/0LDfK10 (accessed on 6 June 2019).

24. International King Tides Project. 2019. Available online: http://kingtides.net/ (accessed on 6 June 2019).

25. Yusuf, J.E.W.; Rawat, P.; Considine, C.; Covi, M.; St. John, B.; Nicula, J.G.; Anuar, K.A. Participatory GIS as a Tool for Stakeholder Engagement in Building Resilience to Sea Level Rise: A Demonstration Project. *Mar. Technol. Soc. J.* **2018**, *52*, 45–55. [CrossRef]

26. Loftis, J.D. Estimating Uncertainty in GPS Measurements from Coastal Citizen Science Monitoring Efforts. *Sensors* **2019**, submitted.

27. Loftis, J.D.; Katragadda, S. Crowdsourcing Hydrocorrection: How Tidewater Virginia Caught the King Tide. In Proceedings of the ESRI GIS User Conference 2018, UC293—Citizen Science at Work, San Diego, CA, USA, 12 July 2018.

28. Sweet, W.V.; Park, J.; Marra, J.J.; Zervas, C.; Gill, S. *Sea Level Rise and Nuisance Flood Frequency Changes Around the United States*; NOAA Technical Report NOS COOPS 73; NOAA: Silver Spring, MD, USA, 2014; 53p.

29. Burgos, A.G.; Hamlington, B.D.; Thompson, P.R.; Ray, R.D. Future Nuisance Flooding in Norfolk, VA, From Astronomical Tides and Annual to Decadal Internal Climate Variability. *Geophys. Res. Lett.* **2018**, *45*, 12–432. [CrossRef]

30. Sweet, W.V.; Marra, J.J. *2015 State of US 'Nuisance' Tidal Flooding*; NOAA Report; NOAA: Silver Spring, MD, USA, 2016.

31. VanHoutven, G.; Depro, B.; Lapidus, D.; Allpress, J.; Lord, B. *Costs of Doing Nothing: Economic Consequences of Not Adapting to Sea Level Rise in the Hampton Roads Region*; Virginia Coastal Policy Center, College of William & Mary Law School Report; Resilient Virginia: Arlington, VA, USA, 2016.

32. Guinness World Records. *The 2019 Guinness Book of World Records*; Record 430293: Most Contributions to an Environmental Survey; Guinness Media: Stamford, CT, USA, 2019; 476p.

33. Loftis, J.D. *Catch the King Tide 2017: All King Tide Data*; College of William & Mary: Williamsburg, VA, USA, 2017. [CrossRef]

34. Loftis, J.D. *Catch the King Tide 2018: All King Tide Data*; College of William & Mary: Williamsburg, VA, USA, 2018. [CrossRef]

35. Loftis, J.D.; Katragadda, S. Validating Operational Flood Forecast Models of King Tides using Citizen Science. In Proceedings of the Esri User Conference, San Diego, CA, USA, 8–12 July 2019; Citizen Scientists Contribute to Data Collection.

36. Zhang, Y.J.; Ye, F.; Stanev, E.V.; Grashorn, S. Seamless cross-scale modeling with SCHISM. *Ocean Model.* **2016**, *102*, 64–81. [CrossRef]

37. Zhang, Y.; Ateljevich, E.; Yu, H.-S.; Wu, C.-H.; Yu, J.C.S. A new vertical coordinate system for a 3D unstructured-grid model. *Ocean Model.* **2015**, *85*, 16–31. [CrossRef]

38. Zhang, Y.; Witter, R.C.; Priest, G.R. Tsunami–tide interaction in 1964 Prince William Sound tsunami. *Ocean Model.* **2011**, *40*, 246–259. [CrossRef]

39. Bertin, X.; Bruneau, N.; Breilh, J.-F.; Fortunato, A.B.; Karpytchev, M. Importance of wave age and resonance in storm surges: The case Xynthia, Bay of Biscay. *Ocean Model.* **2012**, *42*, 16–30. [CrossRef]

40. Rodrigues, M.; Oliveira, A.; Queiroga, H.; Fortunato, A.B.; Zhang, Y. Three-dimensional modeling of the lower trophic levels in the Ria de Aveiro (Portugal). *Ecol. Model.* **2009**, *220*, 1274–1290. [CrossRef]

41. Pinto, L.; Fortunato, A.B.; Zhang, Y.; Oliveira, A.; Sancho, F.E.P. Development and validation of a three-dimensional morphodynamic modelling system. *Ocean Model.* **2012**, *57–58*, 1–14. [CrossRef]

42. Azevedo, A.; Oliveira, A.; Fortunato, A.B.; Zhang, Y.; Baptista, A.M. A cross-scale numerical modeling system for management support of oil spill accidents. *Mar. Pollut. Bull.* **2014**, *80*, 132–147. [CrossRef] [PubMed]

43. Wang, H.V.; Loftis, J.D.; Liu, Z.; Forrest, D.R.; Zhang, Y.J. The storm surge and sub-grid inundation modeling in New York City during Hurricane Sandy. *J. Mar. Sci. Eng.* **2014**, *2*, 226–246. [CrossRef]

44. Yusuf, J.-E.; Considine, C.; Covi, M.; Council, D.; Loftis, J.D. *Preferences for Modeling Scenarios and Parameters: The Perspective of Planners and Emergency Managers*; Paper No. 1 in the Risk Communication and Public Engagement in Sea Level Rise Resilience Research Series; Resilience Collaborative Occasional Paper Series, No. 2017-2; Old Dominion University Resilience Collaborative: Norfolk, VA, USA, 2017.

45. Danielson, J.J.; Poppenga, S.K.; Brock, J.C.; Evans, G.A.; Tyler, D.J.; Gesch, D.B.; Thatcher, C.A.; Barras, J.A. Topobathymetric elevation model development using a new methodology: Coastal national elevation database. *J. Coast. Res.* **2016**, *76*, 75–89. [CrossRef]

46. Loftis, J.D.; Wang, H.V.; DeYoung, R.J.; Ball, W.B. Using Lidar Elevation Data to Develop a Topobathymetric Digital Elevation Model for Sub-Grid Inundation Modeling at Langley Research Center. *J. Coast. Res.* **2016**, *SI76*, 134–148. [CrossRef]

47. Boon, J.D.; Mitchell, M.; Loftis, J.D.; Malmquist, D.M. *Anthropocene Sea Level Change: A History of Recent Trends Observed in the U.S. East, Gulf, and West Coast Regions*; Special Report in Applied Marine Science and Ocean Engineering (SRAMSOE), No. 467; Virginia Institute of Marine Science, William & Mary: Gloucester Point, VA, USA, 2018.

Journal of
Marine Science and Engineering

Article

Domain Decomposition Method for the Variational Assimilation of the Sea Level in a Model of Open Water Areas Hydrodynamics

Valery Agoshkov [1,2], Natalia Lezina [1,*] and Tatiana Sheloput [1]

[1] Marchuk Institute of Numerical Mathematics, Russian Academy of Sciences, 119333 Moscow, Russia; agoshkov@inm.ras.ru (V.A.); sheloput@phystech.edu (T.S.)

[2] Lomonosov Moscow State University, 119991 Moscow, Russia

* Correspondence: lezina@phystech.edu

Received: 30 April 2019; Accepted: 20 June 2019; Published: 23 June 2019

Abstract: One of the modern fields in mathematical modelling of water areas is developing hybrid coastal ocean models based on domain decomposition. In coastal ocean modelling a problem to be solved is setting open boundary conditions. One of the methods dealing with open boundaries is variational data assimilation. The purpose of this work is to apply the domain decomposition method to the variational data assimilation problem. The method to solve the problem of restoring boundary functions at the liquid boundaries for a system of linearized shallow water equations is studied. The problem of determining additional unknowns is considered as an inverse problem and solved using well-known approaches. The methodology based on the theory of optimal control and adjoint equations is used. In the paper the theoretical study of the problem is carried out, unique and dense solvability of the problem is proved, an iterative algorithm is proposed and its convergence is studied. The results of the numerical experiments are presented and discussed.

Keywords: open boundaries; domain decomposition; variational data assimilation; inverse problems; shallow water equations; boundary conditions; mathematical modelling; coastal ocean modelling; computational methods

---

1. Introduction

At the present time, one of the rapidly developing fields is mathematical modelling of water areas of particular interest (seas, bays, open ocean areas) and open coastal regions. The relevance of this topic is justified by the need to assess anthropogenic impacts on marine areas and the consequences of such impacts. Moreover, a number of problems connected to the climate changes for a selected water area over several decades gain increasing interest. In order to simulate coastal ocean processes, involved with various types of physical phenomena occurring at a wide range of spatial and temporal scales, special approaches are required. To provide the correct representation of coastal ocean flows, regional models are created.

Regional ocean modelling has one unavoidable challenge relative to its global counterpart: open boundary conditions (OBC). By definition, a regional ocean model includes open boundaries over at least a part of its perimeter [1]. The "outer liquid" (open) boundary means the "water-to-water" boundary separating the considered area from an ocean. The result obtained both in long-term simulations and in operational forecasting directly depends on the method of setting the OBC. According to [2], in flow problems dominated by advection and/or wave motion, the OBC should allow phenomena generated in the domain of interest and coming from outside to pass through the boundary without undergoing significant distortion and without influencing the interior solution. One of the difficulties in setting such conditions is that there is no accurate information on external energy and mass flows. If open boundaries are located in dynamically active areas, inaccurate accounting for this information leads to inconsistency of the results obtained with the observed fields of physical parameters. For the long-term climate modelling, the appropriate setting of boundary conditions at liquid boundaries is of particular importance.

There are different known methods [3–5] dealing with open boundaries in limited-area models. In a number of studies [6,7] the models are previously set up for a larger area and the results of the preliminary (diagnostic) simulations with these models are used to determine OBC. One-way and two-way grid-nesting techniques are devised to exchange information at the interface between fine-mesh regional models and coarse-mesh large-domain models. The use of averaged data for flows across the open boundary is also acceptable in some cases [8]. A large number of OBC have been proposed in the literature (see [9], for a review). The adaptive algorithm described in [3] is used in many recent studies. It is also possible to use sequential [1] or variational [5] data assimilation methods in order to reduce model-data misfit caused by unsatisfactory OBC. While there are many suggested methods, it is still a question of debate which methodology is suitable for a particular problem.

One of the approaches used for modelling multiphysics and multiscale coastal ocean processes is developing of hybrid models based on domain decomposition. The use of domain decomposition method (DDM) allows to reduce the solution process in the original domain to alternate solving the problem in subdomains, possibly having a simpler form, or apply different models in them (for example, the coupling of first and second order equations [10]). Particularly, it is possible to use DDM to combine models developed for individual phenomena at specific scales. For example, in [11,12] the idea of coupling a geophysical fluid dynamics model and a fully 3D fluid dynamics model in order to simulate multiphysics coastal ocean flows is presented and discussed. Implementation of DDM often requires "inner liquid" boundaries, separating the subdomains. In these subdomains one can obtain the results of simulations using meshes of different scales to achieve better approximation of boundary, bottom topography, etc. At the present time, the development of new algorithms and their effective implementation on multiprocessor computer systems can be attributed to the main goals of the domain decomposition. This makes DDMs be promising in mathematical modelling of processes in the oceans and seas [11–13]. Most studies on application of data assimilation (DA) together with DDM are related to developing parallel algorithms. Some studies [14–17] demonstrate the scalability of the domain decomposition approach and its mathematical consistency when applied in variational DA. Besides, DDM in variational DA problems is suitable not only for creating high-performance algorithms. Particularly, observational data could be available not in the whole modelling area but only in some subdomain in which variational DA procedure may be considered. In those cases the application of DDMs may be effective. This topic is a relatively new field in ocean modelling.

The approach described in this paper is based on [13,18–21]. In [18] the model based on primitive equations written in spherical σ-coordinates with a free surface in the hydrostatic and Boussinesq approximations is considered. For time approximation of the model the splitting method is used [22,23]. For this purpose the whole time interval is divided into subintervals; at each subinterval, the following subproblems named the steps of the splitting method are solved:

Step 1. The heat transfer problem.
Step 2. The salt transport problem.
Step 3. The problem of hydrological fields adaptation. It is solved in 3 substeps: (a) calculation of density and finding corrections to the velocity; (b) solving the problem of baroclinic adaptation; (c) solving the problem of barotropic adaptation (i.e., the system of linearized shallow water equations is considered).

The steps of the splitting method are formulated in [18]. The splitting method allows to consider the DA problem as a sequence of linear DA problems. In [19–21,24,25] an investigation of some of them is given. In [25] an iterative algorithm for solving the problem of variational assimilation of the temperature corresponding to the step 1 of the splitting method is considered. The numerical experiments on the efficiency of the algorithm in the Baltic Sea area are carried out. In [19] an inverse problem of determining an unknown boundary function in the OBC for the simplest model of tides is studied. The results of the work are used in [21], where the algorithm for solving the problem of variational assimilation of the sea level anomaly at the liquid boundary corresponding to step 3 of the splitting method is considered and tested in the Baltic Sea circulation model. In [13] a new approach to formulation of DDM is discussed. This approach was applied to a convection-diffusion problem. The method described in [13] was implemented in the heat transfer block (step 1) of the Baltic Sea dynamics model.

The purpose of this work is to apply DDM to the variational DA problem. The method for solving the problem of restoring boundary functions at the "outer" and "inner" liquid boundaries based on the methods of variational DA and domain decomposition for the subproblem corresponding to a system of linearized shallow water equations (step 3-c) is studied. The problem of determining additional unknowns ("boundary functions") in the boundary conditions is considered as inverse and solved using well-known approaches [26,27]. The major problem in coastal ocean modelling is the reconciliation of model results with observational data. The approach described in the paper may be applied to regional ocean models in order to reduce a model-data misfit and the dependence of the results upon the unsatisfactory OBC. The choice of the approach to formulate the DDM in the present work is justified by a possibility of generalization, since the models in the subdomains may differ. Therefore, the results of this paper may be promising for multiscale and multiphysics coastal processes simulation.

2. Problem Statement

2.1. Notations and Preliminary Notes

Let us consider geographical (geodesic) system of coordinates (λ, θ, r), where $\lambda \in [0, 2\pi]$ is the geographical longitude increasing from West to East, $\theta \in [-\pi/2, \pi/2]$ is the geographical latitude increasing form South to North ($\theta = \phi - \pi/2, \phi \in [0, \pi]$), r is the distance from the center of the Earth to a given point, the mean Earth radius is R_E. Instead of r it is often convenient to introduce the coordinate $z = R_E - r$ of the axis Oz, directed along the normal to the center of the sphere S_R of radius R_E, along the direction of the gravity force. The unit vectors in λ-, θ- and z-directions are denoted by e_λ, e_θ, e_z, respectively. In this case the velocity vector in the ocean is written in the form: $(u_1, u_2, u_3)^T = u_1 e_\lambda + u_2 e_\theta + u_3 e_z \equiv (\vec{u}, w)^T$.

Let Ω denote the connected manifold on the sphere S_R, which is called the "reference surface" [18]. Below we consider the case when Ω does not include polar points. The ocean surface elevation is given by the equation $z = \xi(\lambda, \theta, t)$, $z = H(\lambda, \theta)$ is the bottom topography function, where $(\lambda, \theta, R_E) \in \Omega$, $t \in [0, T]$ is time variable ($T < \infty$), $H(\lambda, \theta) > 0$. Moreover, suppose that there exists $\varepsilon > 0$, so that $H \geq \varepsilon$. Considering these notations, the total depth of the ocean is expressed as $H_{tot\_depth} = H - \xi$

Below we use the following notations of differential operators:

$$\mathbf{grad}\ \Phi = \left(m\frac{\partial \Phi}{\partial \lambda}, n\frac{\partial \Phi}{\partial \theta} \right)^T, \quad \mathbf{div}\ \Phi = m\frac{\partial \Phi}{\partial \lambda} + m\frac{\partial \left[(n/m)\Phi \right]}{\partial \theta},$$

where $m = 1/(R_E \cos\theta)$, $n = 1/R_E$. By $\vec{n}$ we denote the unit vector of outer normal to $\partial\Omega$. Let l denote the Coriolis parameter, g is the gravitational acceleration. Let us introduce the following notation for the depth-averaged horizontal velocities:

$$\vec{U} = \frac{1}{H} \int_\xi^H \vec{u}\, dz.$$

In the current study we consider the subproblem corresponding to a system of linearized shallow water equations (Step 3-c of the splitting scheme introduced above) on the time subinterval (t_{j-1}, t_j), $\Delta t = t_j - t_{j-1}, t_0 = 0, t_J = T, j = 1, \ldots, J$ [18]:

$$
\begin{cases}
\vec{U}_t + \begin{bmatrix} 0 & -\ell \\ \ell & 0 \end{bmatrix} \vec{U} + R_f \vec{U} - g \cdot \mathbf{grad}\, \xi = \vec{f}, & \text{in } \Omega \times (t_{j-1}, t_j), \\
\xi_t - \mathbf{div}\left(H\vec{U} \right) = 0, & \text{in } \Omega \times (t_{j-1}, t_j), \\
\vec{U}(t_{j-1}) = \vec{U}_{j-1}, \quad \xi(t_{j-1}) = \xi_{j-1}, & \text{in } \Omega,
\end{cases}
\tag{1}
$$

where $\vec{f}$ is a given function, R_f is the linear drag coefficient. Solving the problem on (t_{j-1}, t_j), we consider the functions $\vec{U}_{j-1}$, ξ_{j-1} to be known. Detailed description of the notations and simplifications can be found, for example, in [18]. In the present work we consider the boundary condition for System (1) of the form [19]:

$$
H\vec{U} \cdot \vec{n} + m_{op} \sqrt{gH}\xi = m_{op}\sqrt{gH}d_s, \qquad \text{on } \partial\Omega \times (t_{j-1}, t_j),
\tag{2}
$$

where m_{op} is the characteristic function of the "outer liquid" (open) boundary Γ_{op}, i.e., $m_{op} = 1$ if $(\lambda, \theta, R_E) \in \Gamma_{op}$, $m_{op} = 0$ otherwise. If the function d_s is defined, Systems (1) and (2) are well posed. This boundary function will be considered below as an additional unknown.

Considering Systems (1) and (2) on (t_{j-1}, t_j), we introduce the following implicit scheme for time approximation:

$$
\begin{cases}
\dfrac{\vec{U}^j}{\Delta t} + \begin{bmatrix} 0 & -\ell \\ \ell & 0 \end{bmatrix} \vec{U}^j + R_f \vec{U}^j - g \cdot \mathbf{grad}\, \xi^j = \vec{f}^j + \dfrac{\vec{U}^{j-1}}{\Delta t}, & \text{in } \Omega, \\
\dfrac{\xi^j}{\Delta t} - \mathbf{div}\left(H\vec{U}^j \right) = \dfrac{\xi^{j-1}}{\Delta t}, & \text{in } \Omega, \\
H\vec{U}^j \cdot \vec{n} + m_{op}\sqrt{gH}\xi^j = m_{op}\sqrt{gH}d_s^j, & \text{on } \partial\Omega.
\end{cases}
\tag{3}
$$

Hereafter the "semi-discrete" System (3) is the subject of the investigation, so for convenience the indices j will be omitted, i.e.: $\vec{U} \equiv \vec{U}^j, \xi \equiv \xi^j, \ldots$ We also introduce vectors $\tilde{\vec{f}} = (f_1^j + (\vec{U}^{j-1})_1/\Delta t, f_2^j + (\vec{U}^{j-1})_2/\Delta t)^T$, $\tilde{f}_3 = \xi^{j-1}/\Delta t$, $\tilde{\vec{f}} \equiv (\tilde{f}_1, \tilde{f}_2)$. Finally, System (3) can be written in the form:

$$
\begin{cases}
\dfrac{\vec{U}}{\Delta t} + \begin{bmatrix} 0 & -\ell \\ \ell & 0 \end{bmatrix} \vec{U} + R_f \vec{U} - g \cdot \mathbf{grad}\, \xi = \tilde{\vec{f}}, & \text{in } \Omega, \\
\dfrac{\xi}{\Delta t} - \mathbf{div}\left(H\vec{U} \right) = \tilde{f}_3, & \text{in } \Omega, \\
H\vec{U} \cdot \vec{n} + m_{op}\sqrt{gH}\xi = m_{op}\sqrt{gH}d_s, & \text{on } \partial\Omega.
\end{cases}
\tag{4}
$$

2.2. Variational Data Assimilation and the Domain Decomposition Method

We assume that the boundary $\partial\Omega$ of the domain Ω is Lipschitz and piecewise C^2–smooth. We also assume that $\Gamma_{in} \subset \Omega$ is a hypersurface of class C^2 [28] and divides the domain Ω into two subdomains Ω_1 and Ω_2 without overlap, $\Omega \equiv \Gamma_{in} \cup \Omega_1 \cup \Omega_2$ (see Figure 1), $\bar{\Gamma}_{in} \equiv \overline{(\partial\Omega_1 \cap \partial\Omega_2)}$, $\Gamma_{in} \subset \partial\Omega_1$ is open in $\partial\Omega_1$, $\Gamma_{in} \subset \partial\Omega_2$ is open in $\partial\Omega_2$. We suppose that $\partial\Omega_1$ and $\partial\Omega_2$ are Lipschitz and piecewise C^2–smooth. Suppose that Γ_{op} ("outer liquid", open boundary) does not intersect with Γ_{in} ("inner liquid" boundary). Moreover, the essential assumption is that $\rho(\Gamma_{in}, \Gamma_{op}) = \inf\limits_{x \in \Gamma_{in}, y \in \Gamma_{op}} \|x - y\|_{R_2} \gg \sqrt{gH} \cdot \Delta t$.

Below we use the index $i = 1, 2$ to indicate the solutions in subdomains ($i = 1, 2$). The problem (4) can be written for the functions $\vec{U}^{(1)}$, $\xi^{(1)}$ in the subdomain Ω_1, for the functions $\vec{U}^{(2)}$, $\xi^{(2)}$ in the subdomain Ω_2. Suppose $\vec{n}_i$ is the outer normal to the boundary $\partial\Omega_i$ of the domain Ω_i, $i = 1, 2$. We will require the fulfillment of the boundary conditions on the inner liquid boundary Γ_{in}:

$$\zeta^{(1)} = \zeta^{(2)}, \quad HU_n^{(1)} = -HU_n^{(2)}, \tag{5}$$

where $U_n^{(i)} = \vec{U}^{(i)} \cdot \vec{n}_i$, $i = 1, 2$.

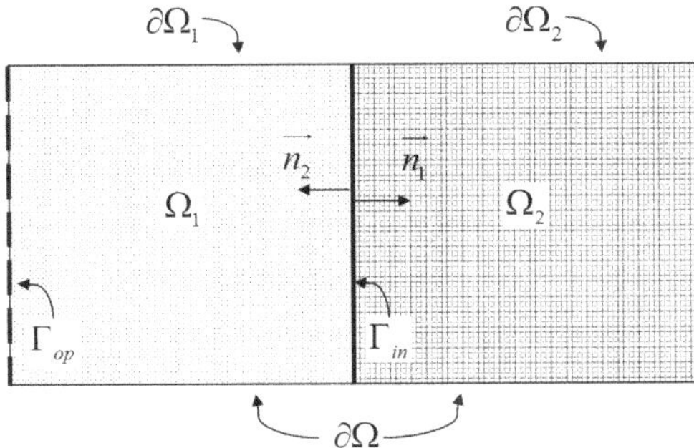

Figure 1. Domain with "inner" Γ_{in} and "outer" Γ_{op} liquid boundaries.

An additional unknown function v is defined as:

$$HU_n^{(1)} = \sqrt{gH}v, \quad \text{on } \Gamma_{in}.$$

We obtain the following form of the boundary conditions for the problem in the subdomain Ω_1:

$$H\vec{U}^{(1)} \cdot \vec{n}_1 + m_{op}\sqrt{gH}\zeta^{(1)} = m_{op}\sqrt{gH}d_s + m_{in}\sqrt{gH}v, \quad \text{on } \partial\Omega_1, \tag{6}$$

and for the problem in the subdomain Ω_2:

$$HU_n^{(2)} = -m_{in}\sqrt{gH}v, \quad \text{on } \partial\Omega_2. \tag{7}$$

Suppose there are preprocessed data of sea level anomaly measurements ζ_{obs} along Γ_{op}. Such data can be obtained from observations or direct measurements (satellite altimetry data, "in-situ" data), or from the simulation using a global model with a coarser grid. In any cases, these data contain errors, so we cannot use ζ_{obs} directly in the OBC. We introduce an additional condition (closure condition):

$$\zeta = \zeta_{obs}, \quad \text{on } \Gamma_{op}. \tag{8}$$

From System (4) in each subdomain ($i = 1, 2$) we obtain:

$$\begin{cases} \dfrac{\vec{U}^{(i)}}{\Delta t} + \begin{bmatrix} 0 & -\ell \\ \ell & 0 \end{bmatrix} \vec{U}^{(i)} + R_f \vec{U}^{(i)} - g \cdot \mathbf{grad}\,\zeta^{(i)} = \tilde{\vec{f}}, \quad \text{in } \Omega_i, \\[2mm] \dfrac{\zeta^{(i)}}{\Delta t} - \mathbf{div}\left(H\vec{U}^{(i)}\right) = \tilde{f}_3, \quad \text{in } \Omega_i, \\[2mm] H\vec{U}^{(1)} \cdot \vec{n}_1 + m_{op}\sqrt{gH}\zeta^{(1)} = m_{op}\sqrt{gH}d_s + m_{in}\sqrt{gH}v, \quad \text{on } \partial\Omega_1, \\[2mm] H\vec{U}^{(2)} \cdot \vec{n}_2 = -m_{in}\sqrt{gH}v, \quad \text{on } \partial\Omega_2. \end{cases} \tag{9}$$

We formulate the *inverse problem of restoring the boundary functions on the "outer" and "inner" liquid boundaries* as follows: find the vector functions of solutions in the subdomains $\Phi^{(i)} = (\vec{U}^{(i)}, \zeta^{(i)})^T$,

$i = 1, 2$, and additional unknown boundary functions v, d_s, satisfying the systems of Equation (9) and additional conditions of Equations (5) and (8).

Consider the Hilbert space $\mathbf{H}_0^{(i)}$ of vector functions $\Phi^{(i)} = (\vec{U}^{(i)}, \xi^{(i)})^T$, $\vec{U}^{(i)} \in (L_2(\Omega_i))^2$, $\xi^{(i)} \in L_2(\Omega_i)$ with the scalar product:

$$(\Phi^{(i)}, \hat{\Phi}^{(i)})_{\mathbf{H}_0^{(i)}} = \int_{\Omega_i} [H(\vec{U}^{(i)} \cdot \hat{\vec{U}}^{(i)}) + g\xi^{(i)}\hat{\xi}^{(i)}] \, d\Omega.$$

By $\mathbf{W}^{(i)}$ we denote a space of vector functions $\Phi \in (L_2(\Omega_i))^2 \times W_2^1(\Omega_i)$. Let $\xi_{obs} \in L_2(\Gamma_{op})$, $\tilde{\vec{f}} \in (L_2(\Omega))^2$, $\tilde{f}_3 \in L_2(\Omega)$, $H \in C^1(\overline{\Omega})$.

The scalar product of (9) and $\hat{\Phi} \in \mathbf{W}^{(i)}$ in $\mathbf{H}_0^{(i)}$, $i = 1, 2$ is:

$$a_1(\Phi^{(1)}, \hat{\Phi}^{(1)}) = f_1(\hat{\Phi}^{(1)}) + b_{op}(d_s, \hat{\Phi}^{(1)}) + b_1(v, \hat{\Phi}^{(1)}), \tag{10}$$

$$a_2(\Phi^{(2)}, \hat{\Phi}^{(2)}) = f_2(\hat{\Phi}^{(2)}) - b_2(v, \hat{\Phi}^{(2)}), \tag{11}$$

where

$$a_1(\Phi^{(1)}, \hat{\Phi}^{(1)}) = \int_{\Omega_1} \left[\frac{H\vec{U}^{(1)}\hat{\vec{U}}^{(1)} + g\xi^{(1)}\hat{\xi}^{(1)}}{\Delta t} + R_f \vec{U}^{(1)}\hat{\vec{U}}^{(1)} + lH((\vec{U}^{(1)})_1(\hat{\vec{U}}^{(1)})_2 - (\vec{U}^{(1)})_2(\hat{\vec{U}}^{(1)})_1) \right] d\Omega +$$
$$+ \int_{\Omega_1} (gH\vec{U}^{(1)} \, \mathbf{grad} \, \hat{\xi}^{(1)} - gH\hat{\vec{U}}^{(1)} \, \mathbf{grad} \, \xi^{(1)}) \, d\Omega + \int_{\Gamma_{op}} g\hat{\xi}^{(1)}\sqrt{gH}\xi^{(1)} \, d\Gamma,$$

$$a_2(\Phi^{(2)}, \hat{\Phi}^{(2)}) = \int_{\Omega_2} \left[\frac{H\vec{U}^{(2)}\hat{\vec{U}}^{(2)} + g\xi^{(2)}\hat{\xi}^{(2)}}{\Delta t} + R_f \vec{U}^{(2)}\hat{\vec{U}}^{(2)} + lH((\vec{U}^{(2)})_1(\hat{\vec{U}}^{(2)})_2 - (\vec{U}^{(2)})_2(\hat{\vec{U}}^{(2)})_1) \right] d\Omega +$$
$$+ \int_{\Omega_2} (gH\vec{U}^{(2)} \, \mathbf{grad} \, \hat{\xi}^{(2)} - gH\hat{\vec{U}}^{(2)} \, \mathbf{grad} \, \xi^{(2)}) \, d\Omega,$$

$$b_{op}(d_s, \hat{\Phi}^{(1)}) = \int_{\Gamma_{op}} \sqrt{gH}d_s g\hat{\xi}^{(1)} \, d\Gamma,$$

$$b_i(v, \hat{\Phi}^{(i)}) = \int_{\Gamma_{in}} \sqrt{gH}vg\hat{\xi}^{(i)} \, d\Gamma, \quad i = 1, 2,$$

$$f_i(\hat{\Phi}^{(i)}) = \int_{\Omega_i} [H\tilde{\vec{f}} \cdot \hat{\vec{U}} + g\tilde{f}_3\hat{\xi}] \, d\Omega, \quad i = 1, 2.$$

Let us introduce the Hilbert space $\mathbf{H}_c$ of vector-functions $u = (d_s, v)^T$, $d_s \in L_2^W(\Gamma_{op})$, $v \in L_2^W(\Gamma_{in})$, with the norm $\|u\|_{\mathbf{H}_c} = \sqrt{\|d_s\|_{L_2^W(\Gamma_{op})}^2 + \|v\|_{L_2^W(\Gamma_{in})}^2}$, where $L_2^W(\Gamma)$ is the space of functions from $L_2(\Gamma)$ with the "weighted" scalar product: $(\cdot, \cdot)_{L_2^W(\Gamma)} = (\sqrt{gH}\cdot, \cdot)_{L_2(\Gamma)}$. We also introduce the space $\mathbf{H}_{ob} = L_2^W(\Gamma_{op}) \times L_2^W(\Gamma_{in})$.

We formulate the problem in a weak form: find $\Phi^{(i)} \in \mathbf{W}^{(i)}$, $u \in \mathbf{H}_c$ satisfying the conditions of Equations (10) and (11) and also the conditions of Equations (5) and (8) (in the sense of equality almost everywhere on Γ_{in}, Γ_{op}, respectively) $\forall \hat{\Phi} \in \mathbf{W}^{(i)}$.

In order to formulate problems in operator form to study the problem theoretically the weak formulation should be modified. To execute this, we use the procedure similar to that described in [19]. From Equation (9) we receive:

$$(\vec{U})_1 = g\frac{\partial \xi}{\partial x}\frac{\vartheta}{\vartheta^2 + l^2} + g\frac{\partial \xi}{\partial y}\frac{l}{\vartheta^2 + l^2} + \tilde{f}_1,$$

$$(\vec{U})_2 = g\frac{\partial \xi}{\partial y}\frac{\vartheta}{\vartheta^2 + l^2} - g\frac{\partial \xi}{\partial x}\frac{l}{\vartheta^2 + l^2} + \tilde{f}_2,$$

where

$$\vartheta = 1/\Delta t + R_f, \quad \tilde{\tilde{f}}_1 = \tilde{f}_1\frac{\vartheta}{\vartheta^2 + l^2} + \tilde{f}_2\frac{l}{\vartheta^2 + l^2}, \quad \tilde{\tilde{f}}_2 = \tilde{f}_2\frac{\vartheta}{\vartheta^2 + l^2} - \tilde{f}_1\frac{l}{\vartheta^2 + l^2}.$$

By matrix M we denote:

$$M = \begin{bmatrix} a & b \\ -b & a \end{bmatrix},$$

where

$$a = \frac{\vartheta}{\vartheta^2 + l^2}, \quad b = \frac{l}{\vartheta^2 + l^2}.$$

Substituting $\vec{U}$ in the weak formulation of the problem we receive:

$$\tilde{a}_1(\xi^{(1)}, \hat{\xi}^{(1)}) = \tilde{f}_1(\hat{\xi}^{(1)}) + \tilde{b}_{op}(d_s, \hat{\xi}^{(1)}) + \tilde{b}_1(v, \hat{\xi}^{(1)}), \tag{12}$$

$$\tilde{a}_2(\xi^{(2)}, \hat{\xi}^{(2)}) = \tilde{f}_2(\hat{\xi}^{(2)}) - \tilde{b}_2(v, \hat{\xi}^{(2)}), \tag{13}$$

where

$$\tilde{a}_1(\xi^{(1)}, \hat{\xi}^{(1)}) = \int_{\Omega_1} \left[\xi^{(1)}\hat{\xi}^{(1)}/\Delta t + gH \cdot M \operatorname{\mathbf{grad}} \xi^{(1)} \cdot \operatorname{\mathbf{grad}} \hat{\xi}^{(1)} \right] d\Omega + \int_{\Gamma_{op}} \hat{\xi}^{(1)} \sqrt{gH}\xi^{(1)} \, d\Gamma,$$

$$\tilde{a}_2(\xi^{(2)}, \hat{\xi}^{(2)}) = \int_{\Omega_2} \left[\xi^{(2)}\hat{\xi}^{(2)}/\Delta t + gH \cdot M \operatorname{\mathbf{grad}} \xi^{(2)} \cdot \operatorname{\mathbf{grad}} \hat{\xi}^{(2)} \right] d\Omega.$$

$$\tilde{b}_{op}(d_s, \hat{\xi}^{(1)}) = \int_{\Gamma_{op}} \sqrt{gH}d_s\hat{\xi}^{(1)} \, d\Gamma, \tag{14}$$

$$\tilde{b}_i(v, \hat{\xi}^{(i)}) = \int_{\Gamma_{in}} \sqrt{gH}v\hat{\xi}^{(i)} \, d\Gamma, \quad i = 1, 2, \tag{15}$$

$$\tilde{f}_i(\hat{\xi}^{(i)}) = \int_{\Omega_i} [\tilde{f}_3\hat{\xi} - HM\tilde{\tilde{f}} \cdot \operatorname{\mathbf{grad}} \hat{\xi}] \, d\Omega, \quad i = 1, 2.$$

Let us introduce the following notation:

$$W_2^1(\Omega_i) \equiv \mathbf{V}^{(i)}$$

The bilinear forms $\tilde{a}_i(\xi^{(i)}, \hat{\xi}^{(i)})$, $i = 1, 2$ are bounded and positive definite for functions $\xi^{(i)}$, $\hat{\xi}^{(i)} \in \mathbf{V}^{(i)}$. Equations (12) and (13) can be formulated in the following form [26]:

$$L_1\xi^{(1)} = \hat{f}_1 + B_{op}d_s + B_1v, \tag{16}$$

$$L_2\xi^{(2)} = \hat{f}_2 - B_2v, \tag{17}$$

where the operators $L_i : \mathbf{V}^{(i)} \to (\mathbf{V}^{(i)})'$, $B_i : L_2^W(\Gamma_{in}) \to (\mathbf{V}^{(i)})'$, $B_{op} : L_2^W(\Gamma_{op}) \to (\mathbf{V}^{(i)})'$ (the space $(\mathbf{V}^{(i)})'$ is the dual of $\mathbf{V}^{(i)}$) are introduced using the bilinear forms $\tilde{a}_i(\xi^{(1)}, \hat{\xi}^{(1)})$, $\tilde{b}_i(v, \hat{\xi}^{(i)})$, $\tilde{b}_{op}(d_s, \hat{\xi}^{(i)})$, $i = 1, 2$, respectively [29], $\hat{f}_i \in (\mathbf{V}^{(i)})'$. The adjoint operators may also be introduced, so that the following identity is satisfied ($i = 1, 2$):

$$\tilde{a}_i(\xi^{(i)}, \hat{\xi}^{(i)}) = (L_i\xi^{(i)}, \hat{\xi}^{(i)}) = (\xi^{(i)}, L_i^*\hat{\xi}^{(i)}). \tag{18}$$

Note, that $(L_i \xi^{(i)}) \in (\mathbf{V}^{(i)})'$, $\hat{\xi}^{(i)} \in \mathbf{V}^{(i)}$ and $(\cdot, \cdot)$ means their scalar product. We obtain the same equalities for the bilinear forms $b_i(v, \hat{\xi}^{(i)})$, $b_{op}(d_s, \hat{\xi}^{(i)})$:

$$b_i(v, \hat{\xi}^{(i)}) = (B_i v, \hat{\xi}^{(i)}) = (v, B_i^* \hat{\xi}^{(i)})_{L_2^W(\Gamma_{in})}, \quad b_{op}(d_s, \hat{\xi}^{(i)}) = (B_{op} d_s, \hat{\xi}^{(i)}) = (d_s, B_{op}^* \hat{\xi}^{(i)})_{L_2^W(\Gamma_{op})}. \tag{19}$$

It could be shown that the operators B_i, B_{op} are bounded, there exist operators L_i^{-1} which are bounded $(i = 1, 2)$ [13,29].

The closure conditions of Equations (5) and (8) are formulated in the form:

$$C_{obs} \xi^{(1)} = \zeta_{obs}, \tag{20}$$

$$C_1 \xi^{(1)} = C_2 \xi^{(2)}, \tag{21}$$

where $C_{obs} \xi = \xi|_{\Gamma_{op}} \, \forall \xi \in \mathbf{V}^{(1)}$ is a trace operator on Γ_{op}, $C_{obs} : \mathbf{V}^{(1)} \to L_2^W(\Gamma_{op})$, $C_i \xi^{(i)} = \xi^{(i)}|_{\Gamma_{in}}$ $\forall \xi^{(i)} \in \mathbf{V}^{(i)}$, $i = 1, 2$, are trace operators on Γ_{in}, $C_i : \mathbf{V}^{(i)} \to L_2^W(\Gamma_{in})$.

Now the *weak formulation of the inverse problem* is equivalent to the following: find $\xi^{(i)}$, d_s, v satisfying Equations (16), (17), (20) and (21).

The system of Equations (16), (17), (20) and (21) can be rewritten in the operator form:

$$Au = \varphi, \tag{22}$$

where

$$A = \begin{bmatrix} C_{obs} L_1^{-1} B_{op} & C_{obs} L_1^{-1} B_1 \\ C_1 L_1^{-1} B_{op} & C_1 L_1^{-1} B_1 + C_2 L_2^{-1} B_2 \end{bmatrix},$$

$$\varphi = \begin{bmatrix} \zeta_{obs} - C_{obs} L_1^{-1} \hat{f}_1 \\ C_2 L_2^{-1} \hat{f}_2 - C_1 L_1^{-1} \hat{f}_1 \end{bmatrix}.$$

Finally, the original problem is reduced to a single operator Equation (22), for which many results of the general theory of operator equations [26,27,30] are applicable. In the current study the methodology based on the methods of optimal control and adjoint equations is used [26].

2.3. Optimal Control Problem

Note that the operator A is bounded: $\|Au\|_{\mathbf{H}_{ob}} \leq c_A \|u\|_{\mathbf{H}_c}$, $c_A = const < \infty$. However Equation (22) may be ill-posed, since the non-smooth observational data $\zeta_{obs} \in L_2(\Gamma_{op})$ are used in setting the function φ, so $\varphi \notin R(A)$. In this regard let us move on to a *generalized formulation of Equation* (22): find $u \in \mathbf{H}_c$ minimizing $J_0(u) = 1/2\|Au - \varphi\|_{\mathbf{H}_{ob}}^2$.

We formulate the class of optimal control problems [26]: find boundary functions $u = (d_t, v)^T \in \mathbf{H}_c$, minimizing the functional $J_\alpha(u)$:

$$J_\alpha(u) = \frac{1}{2}\alpha\|u\|_{\mathbf{H}_c}^2 + \frac{1}{2}\|Au - \varphi\|_{\mathbf{H}_{ob}}^2, \; \alpha \geq 0. \tag{23}$$

Note that the necessary condition for the minimum of the functional $J_\alpha(u)$ has the form of an equation in terms of Tikhonov regularization method, where α is a regularization parameter. Henceforth, we will call α the *regularization parameter*. If $\alpha > 0$, functional defined in Equation (23) is strictly convex, strongly convex and has a unique global minimum $J^* = J_\alpha^*(u^*(\alpha))$. If $\varphi \in R(A)$ and the solution of the inverse problem is unique, $u^*(\alpha)$ tends to the solution when $\alpha \to +0$. Note also that for $\alpha = 0$ the optimal control problem is equivalent to the generalized formulation of Equation (22).

3. Uniqueness of the Solution

Let the kernel of the operator A consist of not only a single zero element. In this case, there exists $u = (d_s, v)^T \neq 0$ and $Au = 0$. This condition is equivalent to the existence of a nonzero weak solution of the problem:

$$
\begin{cases}
\dfrac{\tilde{\vec{U}}^{(i)}}{\Delta t} + \begin{bmatrix} 0 & -\ell \\ \ell & 0 \end{bmatrix} \tilde{\vec{U}}^{(i)} + R_f \tilde{\vec{U}}^{(i)} - g \cdot \mathbf{grad}\, \tilde{\xi}^{(i)} = 0, & \text{in } \Omega_i, \\[3mm]
\dfrac{\tilde{\xi}^{(i)}}{\Delta t} - \mathbf{div}\left(H \tilde{\vec{U}}^{(i)} \right) = 0, & \text{in } \Omega_i, \\[3mm]
H \tilde{\vec{U}}^{(1)} \cdot \vec{n}_1 + m_{op}\sqrt{gH}\tilde{\xi}^{(1)} = m_{op}\sqrt{gH}d_s + m_{in}\sqrt{gH}v, & \text{on } \partial\Omega_1, \\[3mm]
\tilde{\xi}^{(1)} = 0, & \text{on } \Gamma_{op}, \\[3mm]
H \tilde{\vec{U}}^{(2)} \cdot \vec{n}_2 = -m_{in}\sqrt{gH}v, & \text{on } \partial\Omega_2,
\end{cases}
$$

where $i = 1, 2$. Note that the functions $\tilde{\xi} = \{\tilde{\xi}^{(i)} \text{ in } \Omega_i, \ i = 1, 2\}$, $\tilde{\vec{U}} = \{\tilde{\vec{U}}^{(i)} \text{ in } \Omega_i, \ i = 1, 2\}$ are the solution of a homogenous boundary-value problem in Ω with mixed boundary conditions. So, we obtain: $\tilde{\xi}^{(i)} = 0$, $\tilde{\vec{U}}^{(i)} = 0$ and $d_s = 0$, $v = 0$. Hence, $u = 0$ and $ker(A) = \{0\}$, i.e., *the problem considered may have only a unique solution.*

4. Optimality Condition

The necessary optimality condition for the functional J_α can be written in the form:

$$
\alpha u + A^* A u = A^* \varphi, \tag{24}
$$

where A^* is the adjoint to A:

$$
A^* = \begin{bmatrix} B_{op}^* L_1^{*-1} C_{obs}^* & B_{op}^* L_1^{*-1} C_1^* \\ B_1^* L_1^{*-1} C_{obs}^* & B_1^* L_1^{*-1} C_1^* + B_2^* L_2^{*-1} C_2^* \end{bmatrix}.
$$

Introduce the following adjoint problem:

$$
L_1^* q_1 = C_{obs}^*(C_{obs}\xi^{(1)} - \xi_{obs}) + C_1^*(C_1\xi^{(1)} - C_2\xi^{(2)}), \tag{25}
$$
$$
L_2^* q_2 = C_2^*(C_1\xi^{(1)} - C_2\xi^{(2)}). \tag{26}
$$

The optimality Equation (24) takes the form:

$$
\begin{cases}
\alpha d_s + B_{op}^* q_1 = 0, \\
\alpha v + B_1^* q_1 + B_2^* q_2 = 0.
\end{cases} \tag{27}
$$

We compute the scalar product of vector function $\hat{u} = (\hat{d}_s, \hat{v})^T \in \mathbf{H}_c$ and Equation (27) in $\mathbf{H}_c$. Using Equations (14) and (15) and representations of the bilinear forms of Equation (19), we receive the integral analogue of the optimality Equation (27):

$$
\alpha \int\limits_{\Gamma_{op}} \sqrt{gH}d_s\hat{d}_s\, d\Gamma + \int\limits_{\Gamma_{op}} \sqrt{gH}q_1\hat{d}_s\, d\Gamma \ +
$$
$$
+\alpha \int\limits_{\Gamma_{in}} \sqrt{gH}v\hat{v}\, d\Gamma + \int\limits_{\Gamma_{in}} \sqrt{gH}(q_1 + q_2)\hat{v} = 0 \quad \forall \hat{u} = (\hat{d}_s, \hat{v})^T \in \mathbf{H}_c.
$$

Similarly, integral relations for adjoint Equations (25) and (26) can be obtained.

Adjoint Equations (25) and (26) in differential form ($\xi^{*,(1)} = q_1$, $\xi^{*,(2)} = q_2$) are given by:

$$
\begin{cases}
\dfrac{\vec{U}^{*,(i)}}{\Delta t} - \begin{bmatrix} 0 & -\ell \\ \ell & 0 \end{bmatrix} \vec{U}^{*,(i)} + R_f \vec{U}^{*,(i)} + g \cdot \mathbf{grad}\, \xi^{*,(i)} = 0, & \text{in } \Omega_i, \\[2ex]
\dfrac{\xi^{*,(i)}}{\Delta t} + \mathbf{div}\left(H\vec{U}^{*,(i)} \right) = 0, & \text{in } \Omega_i, \\[2ex]
- H\vec{U}^{*,(1)} \cdot \vec{n}_1 + m_{op}\sqrt{gH}\xi^{*,(1)} = m_{op}\sqrt{gH}(\xi^{(1)} - \xi_{obs}) + m_{in}\sqrt{gH}(\xi^{(1)} - \xi^{(2)}), & \text{on } \partial\Omega_1, \\[2ex]
- H\vec{U}^{*,(2)} \cdot \vec{n}_2 = m_{in}\sqrt{gH}(\xi^{(1)} - \xi^{(2)}), & \text{on } \partial\Omega_2.
\end{cases}
\tag{28}
$$

The optimality conditions take the form:

$$
\alpha d_s + \xi^{*,(1)} = 0, \quad \text{on } \Gamma_{op},
\tag{29}
$$

$$
\alpha v + (\xi^{*,(1)} + \xi^{*,(2)}) = 0, \quad \text{on } \Gamma_{in}.
\tag{30}
$$

So, the functions d_s, v minimizing J_α, $\alpha \geq 0$, satisfy the optimality Equations (29) and (30), where $\xi^{*,(i)}$, $i = 1, 2$ are the weak solutions of Equation (28), $i = 1, 2$, in which $\xi^{(i)}$ are the weak solutions of the systems of Equation (9).

5. Dense Solvability

Now we consider the following problems:

$$
\tilde{a}_1(\hat{\xi}^{(1)}, \xi^{*,(1)}) = \int_{\Gamma_{op}} \sqrt{gH}w\hat{\xi}^{(1)}\, d\Gamma + \int_{\Gamma_{in}} \sqrt{gH}h\hat{\xi}^{(1)}\, d\Gamma, \quad \forall \hat{\xi}^{(1)} \in \mathbf{V}^{(1)}
$$

$$
\tilde{a}_2(\hat{\xi}^{(2)}, \xi^{*,(2)}) = \int_{\Gamma_{in}} \sqrt{gH}h\hat{\xi}^{(2)}\, d\Gamma, \quad \forall \hat{\xi}^{(2)} \in \mathbf{V}^{(2)}
$$

with some (possibly non-trivial) functions $h \in L_2(\Gamma_{in})$, $w \in L_2(\Gamma_{op})$. Let the functions $\xi^{*,(1)}$, $\xi^{*,(2)}$ satisfy additional conditions

$$
\xi^{*,(1)} = 0, \quad \text{on } \Gamma_{op},
$$

$$
\xi^{*,(1)} + \xi^{*,(2)} = 0, \quad \text{on } \Gamma_{in}.
$$

Note that the functions $\xi^* = \xi^{*,(i)}$ in Ω_i, $i = 1, 2$, are the solution of a homogenous boundary-value problem in Ω [13] and $\xi^* = 0$ in Ω. So, we obtain: $\xi^{*,(i)} = 0$ in Ω_i, $i = 1, 2$ and $w = 0$, $h = 0$. Hence, $ker(A^*) = \{0\}$, and this means the *dense solvability* of the problem [26].

6. Iterative Algorithm

As it was shown in the Sections 3 and 5, the problem to find $\Phi^{(i)} = (\vec{U}^{(i)}, \xi^{(i)})^T$, $i = 1, 2$ and the additional boundary functions v, d_s is uniquely and densely solvable. Therefore the functions $\Phi^{(i)}(\alpha)$, $v(\alpha)$, $d_s(\alpha)$ satisfying Equations (9) and (28)–(30) could be taken as an approximation to the solution of the original problem [26]. An approximation to $\Phi^{(i)}(\alpha)$, $v(\alpha)$, $d_s(\alpha)$ could be found with the following iterative algorithm.

1. Let $u^k = (d_s^k, v^k)^T$ be found. Solve the problems in each subdomain Ω_i, $i = 1, 2$:

$$
\begin{cases}
\dfrac{\vec{U}^{(i),k}}{\Delta t} + \begin{bmatrix} 0 & -\ell \\ \ell & 0 \end{bmatrix} \vec{U}^{(i),k} + R_f \vec{U}^{(i),k} - g \cdot \mathbf{grad}\, \xi^{(i),k} = \vec{\tilde{f}}, & \text{in } \Omega_i, \\[2mm]
\dfrac{\xi^{(i),k}}{\Delta t} - \mathbf{div}\left(H\vec{U}^{(i),k} \right) = \tilde{f}_3, & \text{in } \Omega_i, \\[2mm]
H\vec{U}^{(1),k} \cdot \vec{n}_1 + m_{op}\sqrt{gH}\xi^{(1),k} = m_{op}\sqrt{gH}d_s^k + m_{in}\sqrt{gH}v^k, & \text{on } \partial\Omega_1, \\[2mm]
H\vec{U}^{(2),k} \cdot \vec{n}_2 = -m_{in}\sqrt{gH}v^k, & \text{on } \partial\Omega_2.
\end{cases}
\tag{31}
$$

2. Solve the adjoint problems in Ω_i, $i = 1, 2$:

$$
\begin{cases}
\dfrac{\vec{U}^{*,(i),k}}{\Delta t} - \begin{bmatrix} 0 & -\ell \\ \ell & 0 \end{bmatrix} \vec{U}^{*,(i),k} + R_f \vec{U}^{*,(i),k} + g \cdot \mathbf{grad}\, \xi^{*,(i),k} = 0, & \text{in } \Omega_i, \\[2mm]
\dfrac{\xi^{*,(i),k}}{\Delta t} + \mathbf{div}\left(H\vec{U}^{*,(i),k} \right) = 0, & \text{in } \Omega_i, \\[2mm]
- H\vec{U}^{*,(1),k} \cdot \vec{n}_1 + m_{op}\sqrt{gH}\xi^{*,(1),k} = m_{op}\sqrt{gH}(\xi^{(1),k} - \xi_{obs}) + m_{in}\sqrt{gH}(\xi^{(1),k} - \xi^{(2),k}), & \text{on } \partial\Omega_1, \\[2mm]
- H\vec{U}^{*,(2),k} \cdot \vec{n}_2 = m_{in}\sqrt{gH}(\xi^{(1),k} - \xi^{(2),k}), & \text{on } \partial\Omega_2.
\end{cases}
\tag{32}
$$

3. Find new $u^{k+1} = (d_s^{k+1}, v^{k+1})^T$ by:

$$
d_s^{k+1} = d_s^k - \tau_k(\alpha d_s^k + \xi^{*,(1),k}), \qquad \text{on } \Gamma_{op},
\tag{33}
$$

$$
v^{k+1} = v^k - \tau_k(\alpha v^k + (\xi^{*,(1),k} + \xi^{*,(2),k})), \qquad \text{on } \Gamma_{in}.
\tag{34}
$$

Using assertions of [26] we obtain the validity of the following theorem:

Theorem 1. *(I) If $\xi_{obs}^{(0)}$ — exact data, ξ_{obs} — measured data of observations (possibly containing errors), $\|\xi_{obs} - \xi_{obs}^{(0)}\|_{L_2(\Gamma_{op})} \leq \delta_{obs}$, $\delta_{obs} > 0$, $\xi^{(1)}(\alpha)$ and $\xi^{(2)}(\alpha)$ are the solutions of the optimality systems of Equations (9) and (28)–(30), $i = 1, 2$, then the following assessment is valid:*

$$
\left(\|\xi^{(1)}(\alpha) - \xi_{obs}\|^2_{L_2^W(\Gamma_{op})} + \|\xi^{(1)}(\alpha) - \xi^{(2)}(\alpha)\|^2_{L_2^W(\Gamma_{in})} \right)^{1/2} \leq c_\alpha \sqrt{\alpha} + c_\delta \delta_{obs},
\tag{35}
$$

where $c_\alpha = const > 0$, $c_\delta = const > 0$.
(II) If $\varphi \in R(A)$, then the inverse problem of Equation (22) has a unique normal solution $u_0 = (d_{s,0}, v_0)^T$. In that case for enough small $\tau_k = \tau > 0$ vector function $u^k(\alpha) = (d_s^k(\alpha), v^k(\alpha))^T$ tends to u_0 in $\mathbf{H}_c$, when $k \to \infty$, $\alpha \to +0$.

When the stopping criterion of iterative algorithm is satisfied, the functions d_s^k, v^k, $\xi^{(i),k}$, $\vec{U}^{(i),k}$ are taken as an approximate solution of the considered Equations (5), (8) and (9) in Ω_i, $i = 1, 2$. Suitable stopping criterion should be chosen depending on the iterative parameters of the algorithm, size of the modelled region, numerical method, and a given accuracy. More often it is chosen as a limitation of iterations or of residual value. Equations (31)–(34) converge for the small enough parameter $\tau_k = \tau = const > 0$. However, τ_k may be chosen as follows [26]:

$$
\tau_k = \frac{J_\alpha(u_\alpha^k)}{\|J'_\alpha(u_\alpha^k)\|^2} = \frac{1}{2} \frac{\int\limits_{\Gamma_{op}} \sqrt{gH}(\xi^{(1)} - \xi_{obs})^2\, d\Gamma + \int\limits_{\Gamma_{in}} \sqrt{gH}(\xi^{(1)} - \xi^{(2)})^2\, d\Gamma}{\int\limits_{\Gamma_{op}} \sqrt{gH}(\xi^{*,(1)})^2\, d\Gamma + \int\limits_{\Gamma_{in}} \sqrt{gH}(\xi^{*,(1)} + \xi^{*,(2)})^2\, d\Gamma}.
\tag{36}
$$

This choice of the parameter τ_k could be helpful to reduce the number of iterations required.

7. Numerical Experiments and Discussion

The described approach to domain decomposition and variational DA is applied to the system of shallow water systems of Equation (4). Here Ω is a domain on the plane, and (x, y) are Cartesian coordinates.

To set initial conditions, a preliminary calculation without the domain decomposition and variational DA methods is carried out. In this case the domain is represented by the rectangular $[-L, L] \times [0, L]$ with $L = 100$ m. The value of the physical parameters are $g = 9.81$ m/s^2, $l = 0$, $R_f = 0$ and $H = -0.7x/L + 1$ (m). The forces $\tilde{\vec{f}}$ and $(\vec{F})_3$ are equal to 0. The boundary conditions are $H(\vec{U} \cdot \vec{n}) = 0$. For this preliminary problem the conditions at $t = 0$ are given by (p is a subscript for the preliminary results):

$$u_p = 0, \ v_p = 0,$$

$$\zeta_p = 10 \exp\left(-\frac{(x - 25)^2 + (y - 30)^2}{100}\right).$$

To discretize in time, an implicit scheme is used. The time step is 0.5 s. The finite difference method with the first order space discretization is applied. The spatial grid is uniform, the spatial grid steps are 1 m. The results of simulation after 25 s are taken as initial conditions to the next series of numerical experiments.

To set the observational data on Γ_{op}, the preliminary experiment is continued on the next 5 s. At each time step after the first 25 s the observational data are chosen as $\zeta_{obs} = \zeta_p(0, y) \cdot (1 + 0.1a - 0.1b)$, where a and b are random numbers in the range $[0, 1)$. The numbers are received by pseudo-random generators for the uniform distribution law. Thus, the obtained observational data are artificially noisy (noise level is 0.1).

So, we get the initial conditions and the observational data for the next experiments with previously described approach of application of DA and DDMs.

For the numerical experiments the domain Ω is the square $[0, L] \times [0, L]$. It is decomposed into the two subdomains $\Omega_1 = [0, L/2] \times [0, L]$ and $\Omega_2 = [L/2, L] \times [0, L]$ without overlap. We denote the "inner" boundary by $\Gamma_{in} = L/2 \times [0, L]$ and the "outer" boundary by $\Gamma_{op} = 0 \times [0, L]$. Here $L = 100$ m. The value of the physical parameters are $g = 9.81$ m/s^2, $l = 0$, $R_f = 0$. The depth of the modelling domain is $H = -0.7x/L + 1$ (m). The forces $\tilde{\vec{f}}$ and $(\vec{F})_3$ are equal to 0.

To discretize in time, an implicit scheme is chosen. The time step is 0.5 s. The finite difference method with the first order space discretization is applied. The spatial grid is uniform, with grid steps 1 m. The described algorithm of domain decomposition and variational DA is implemented at each time step.

Table 1 gives the residual value depending on the noise level and the stopping criterion (the number of iterations less than 10 or less than 50). The residual value here is defined by

$$Res = \left(\int_{\Gamma_{op}} \sqrt{gH}(\zeta^{(1)} - \zeta_{obs})^2 \, d\Gamma + \int_{\Gamma_{in}} \sqrt{gH}(\zeta^{(1)} - \zeta^{(2)})^2 \, d\Gamma\right)^{1/2}$$

From Table 1 one can see that the noise level has a significant impact on the residual value at the 50th iteration and has almost no affect at the 10th iteration.

Table 1. The residual value (*Res*) depending on the stopping criterion and noise level

Noise Level	0	0.05	0.1
10 iterations	3.65×10^{-1}	3.66×10^{-1}	3.68×10^{-1}
50 iterations	8.80×10^{-5}	5.58×10^{-3}	1.47×10^{-2}

Further in this section the stopping criterion is the number of iterations less than 50. Figure 2 shows the results of modelling for the preliminary problem and the results of modelling using domain decomposition and variational DA. The "inner" boundary is presented as a white line. As seen in Figure 2a,b, the results differ from each other. It is connected with the application of variational DA on the open boundary, because the observational data ξ_{obs} used as closure condition Equation (8) have some additional noise and the results reproduce it (as can be seen in Figure 3).

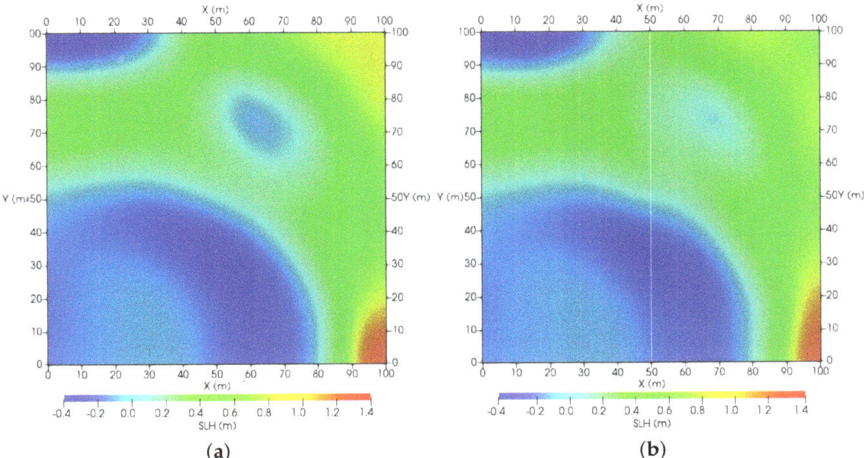

Figure 2. Sea level function (m) at $t = 5$ s after the preliminary simulation: (**a**) Part of the preliminary calculation. (**b**) Simulation with domain decomposition and data assimilation (DA) methods.

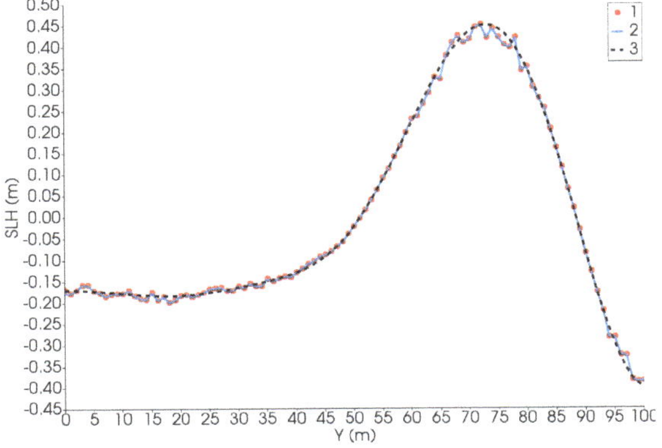

Figure 3. Sea level function on Γ_{op} and observational data (1—observational data; 2—results of simulation; 3—results of the preliminary calculation).

The iterative algorithm converges with the chosen parameter τ_k from Equation (36). Figure 4 shows the results of sea level functions at the first, the ninth and the last iterations. For clarity, the observational data are also presented.

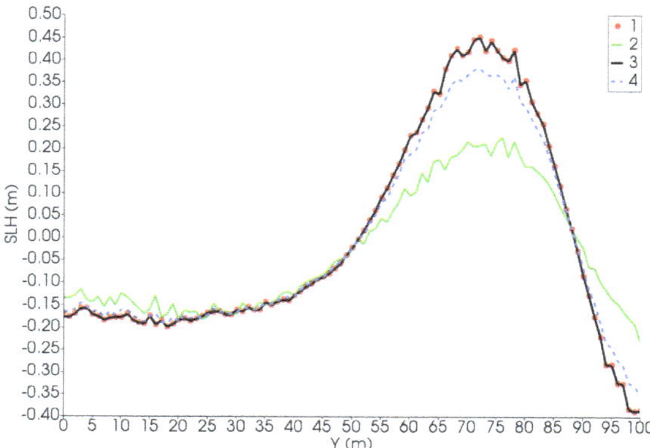

Figure 4. Sea level function on Γ_{op} at different iterations and observational data (1—observational data; 2—the first iteration; 3—the last iteration; 4—the ninth iteration).

In order to analyse the impact on the simulation results of the DDM, the comparison of sea level function results at the first and the last iterations on Γ_{in} is presented in Figure 5. We could note that $\xi^{(1)}$ almost coincides with $\xi^{(2)}$ on the "inner" boundary at the last iteration.

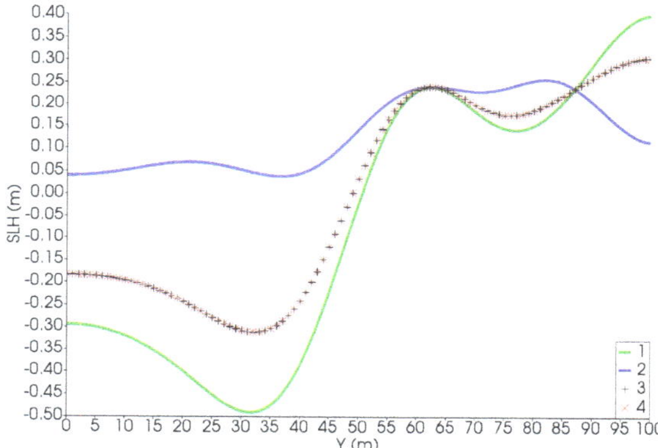

Figure 5. Sea level function on Γ_{in} at different iterations (1—$\xi^{(1)}$ at the first iteration; 2—$\xi^{(2)}$ at the first iteration; 3—$\xi^{(1)}$ at the last iteration; 4—$\xi^{(2)}$ at the last iteration).

In addition, we study the dependence on the regularization parameter α. According to the theory [27], an increase in the regularization parameter α leads to a decrease in the influence of additional noise on the solution. Table 2 provides the error

$$\|\xi^{(1)} - \xi_p\|_{L_2^W(\Gamma_{op})} = \left(\int_{\Gamma_{op}} \sqrt{gH}\, (\xi - \xi_p)^2 \, d\Gamma \right)^{1/2}$$

depending on the regularization parameter α and the stopping criterion (the number of iterations less than 10 or less than 50). As expected, the error increases at the beginning, because the accuracy of the method is not sufficient to reproduce the additional noise (see Figure 4 and Table 1). When the stopping criterion is the number of iterations less than 50, the regularization parameter smooth the result.

Table 2. The error $\|\xi^{(1)} - \xi_p\|_{L_2^W(\Gamma_{op})}$ depending on the stopping criterion and regularization parameter α.

α	10^{-5}	10^{-4}	10^{-3}	10^{-2}
10 iterations	3.52×10^{-1}	3.51×10^{-1}	3.71×10^{-1}	3.92×10^{-1}
50 iterations	1.70×10^{-1}	1.67×10^{-1}	1.60×10^{-1}	1.41×10^{-1}

Thus, application of DDMs in variational DA problem is considered. The time required to solve DA problem with using DDM slightly increases (by 5%) in contrast to one required to solve only DA procedure. The developing of a parallel algorithm based on the described approach may decrease the computation time.

8. Conclusions

The algorithm of DDM in the problem of variational DA is considered. The theoretical study of the problem has been carried out, including the study of the unique and dense solvability of the inverse problem. The iterative algorithm is proposed and the theorem concerning its convergence has been formulated. To illustrate the theoretical results the numerical experiments for the linearized shallow water equations have been carried out. The results of the numerical experiments show that the iterative algorithm converges with the parameter τ_k described in this paper. The DDM does not significantly affect the modelling results. The result of the simulation with variational DA method fits the observational data. However, the additional noise in the observational data yields a noise in the solution results. Experiments show that the effect of a noise may be mitigated by the regularisation. It is worth to be noted that the regularisation parameter and the iteration parameter should be chosen depending on a problem to be solved. These parameters affect the accuracy, convergence and rate of convergence.

In this paper we focus on the theoretical study of the inverse problem for the linearized system of shallow water equations. To illustrate the theoretical results the simplified example was considered. To estimate the approach, its quality and the possibility of application to realistic problems, more experimental results should be provided. We are going to continue the study and test the algorithm in the regional ocean models (e.g., [22,31]). In this case the algorithm will be implemented at each time step to the problem corresponding to step 3-c of the splitting method (mentioned in Section 1). The general approach outlined here may be extended to other regional ocean models. However, for each special problem, boundary conditions on the inner and outer liquid boundaries should be formulated depending on chosen simplifications and the corresponding theoretical study should be carried out. It is worth noting that the methodology based on the theory of optimal control and adjoint equations (used in this paper) may be applied to nonlinear problems. In addition, it may become possible to use meshes of different scales in subdomains. Moreover, the general approach may be improved to become suitable for modelling multiphysics and multiscale coastal ocean processes.

Author Contributions: V.A. stated the problem and suggested the methodology; N.L. and T.S. investigated the problem, proposed an algorithm, carried out the numerical experiments; V.A., N.L. and T.S. analysed the experiments.

Funding: The work was supported by the Russian Science Foundation (project 19–71–20035, studies in Sections 2 and 7) and by the Russian Foundation for Basic Research (project 18-31-00096 mol_a, studies in Sections 3–6).

Conflicts of Interest: The authors declare no conflict of interest.

References

1. Edwards, C.A.; Moore, A.M.; Hoteit, I.; Cornuelle, B.D. Regional Ocean Data Assimilation. *Annu. Rev. Mar. Sci.* **2015**, *7*, 1–22. [CrossRef] [PubMed]

2. Orlanski, I. A simple boundary condition for unbounded hyperbolic flows. *J. Comput. Phys.* **1976**, *21*, 251–269. [CrossRef]

3. Marchesiello, P.; McWilliams, J.C.; Shchepetkin, A. Open boundary conditions for long-term integration of regional oceanic models. *Ocean Model.* **2001**, *3*, 1–20. [CrossRef]

4. Chapman, D.C. Numerical treatment of cross-shelf open boundaries in a barotropic coastal ocean model. *J. Phys. Oceanogr.* **1985**, *15*, 1060–1075. [CrossRef]

5. Bennett, A.F.; McIntosh, P.C. Open ocean modelling as an inverse problem: tidal theory. *J. Phys. Oceanogr.* **1982**, *12*, 1004–1018. [CrossRef]

6. Budgell, W.P. Numerical simulation of ice-ocean variability in the Barents Sea region. *Ocean Dyn.* **2005**, *55*, 370–387. [CrossRef]

7. Oddo, P.; Pinardi, N. Lateral open boundary conditions for nested limited area models: A scale selective approach. *Ocean Model.* **2008**, *20*, 134–156. [CrossRef]

8. Chernov, I.; Tolstikov, A. Comparing two models of large-scale White sea hydrodynamics and thermal dynamics. In Proceedings of the 9th International Scientific and Practical Conference, Rezekne, Latvia, 20–22 June 2013.

9. Palma, E.D.; Matano, R.P. On the implementation of passive open boundary conditions for a general circulation model: The barotropic mode. *J. Geophys. Res.* **1998**, *103*, 1319–1341. [CrossRef]

10. Agoshkov, V.; Gervasio, P.; Quarteroni, A. Optimal control in heterogeneous domain decomposition methods for advection-diffusion equations. *Mediterr. J. Math.* **2006**, *3*, 147–176. [CrossRef]

11. Tang, H.S.; Qu, K.; Wu, X.G. An overset grid method for integration of fully 3D fluid dynamics and geophysics fluid dynamics models to simulate multiphysics coastal ocean flows. *J. Comput. Phys.* **2014**, *273*, 548–571. [CrossRef]

12. Qu, K.; Tang, H.S.; Agrawal, A.; Jiang, C.B.; Deng, B. Evaluation of SIFOM-FVCOM system for high-fidelity simulation of small-scale coastal ocean flows. *J. Hydrodyn. Ser. B* **2016**, *28*, 994–1002. [CrossRef]

13. Agoshkov, V.I. *Domain Decomposition Methods in Problems of Oceans and Seas Hydrothermodynamics*; INM RAS: Moscow, Russia, 2017; pp. 99–125. (In Russian)

14. Sheloput, T.O.; Lezina, N.R. Joint realization of the methods of data assimilation on 'liquid' boundaries and domain decomposition in the Baltic Sea. *Vestnik Tverskogo Gosudarstvennogo Universiteta, Seriya: Geografiya i Geoekologiya* **2018**, *3*, 168–179.

15. D'Amore, L.; Arcucci, R.; Carracciuolo, L.; Murli, A. DD-OceanVar: A Domain Decomposition Fully Parallel Data Assimilation Software for the Mediterranean Forecasting System. *Procedia Comput. Sci.* **2013**, *18*, 1235–1244. [CrossRef]

16. Teruzzi, A.; Di Cerbo, P.; Cossarini, G.; Pascolo, E.; Salon, S. Parallel implementation of a data assimilation scheme for operational oceanography: The case of the MedBFM model system. *Comput. Geosci.* **2019**, *124*, 103–114. [CrossRef]

17. Trémolet, Y.; Le Dimet, F.X. Parallel algorithms for variational data assimilation and coupling models. *Parallel Comput.* **1996**, *22*, 657–674. [CrossRef]

18. Agoshkov, V.I. Statement and study of some inverse problems in modelling of hydrophysical fields for water areas with 'liquid' boundaries. *Russ. J. Numer. Anal. Math. Model.* **2017**, *32*, 73–90. [CrossRef]

19. Agoshkov, V.I. Inverse problems of the mathematical theory of tides: Boundary-function problem. *Russ. J. Numer. Anal. Math. Model.* **2005**, *20*, 1–18. [CrossRef]

20. Dement'eva, E.V.; Karepova, E.D.; Shaidurov, V.V. Assimilation of observation data in the problem of surface wave propagation in a water area with an open boundary. *Russ. J. Numer. Anal. Math. Model.* **2014**, *29*, 13–23. [CrossRef]

21. Sheloput, T.O. Numerical solution of the problem of variational assimilation of the sea level on the liquid (open) boundary in the Baltic Sea hydrothermodynamics model. *Sovremennye Problemy Distantsionnogo Zondirovaniya Zemli iz Kosmosa* **2018**, *15*, 15–23. [CrossRef]

22. Zalesny, V.B.; Gusev, A.V.; Ivchenko, V.O.; Tamsalu, R.; Aps, R. Numerical model of the Baltic Sea circulation. *Russ. J. Numer. Anal. Math. Model.* **2013**, *28*, 85–100. [CrossRef]

23. Zalesny, V.B.; Marchuk, G.I.; Agoshkov, V.I.; Bagno, A.V.; Gusev, A.V.; Diansky, N.A.; Moshonkin, S.N.; Tamsalu, R.; Volodin, E.M. Numerical simulation of large-scale ocean circulation based on the multicomponent splitting method. *Russ. J. Numer. Anal. Math. Model.* **2010**, *25*, 581–609. [CrossRef]

24. Agoshkov, V.I.; Sheloput, T.O. The study and numerical solution of the problem of heat and salinity transfer assuming 'liquid' boundaries. *Russ. J. Numer. Anal. Math. Model.* **2016**, *31*, 71–80. [CrossRef]

25. Agoshkov, V.I.; Sheloput, T.O. The study and numerical solution of some inverse problems in simulation of hydrophysical fields in water areas with 'liquid' boundaries. *Russ. J. Numer. Anal. Math. Model.* **2017**, *32*, 147–164. [CrossRef]

26. Marchuk, G.I.; Agoshkov, V.I.; Shutyaev, V.P. *Adjoint Equations and Perturbation Algorithms in Nonlinear Problems*; CRC Press: Boca Raton, FL, USA, 1996; pp. 1–288.

27. Tikhonov, A.N.; Arsenin, V.Y. *Solution of Ill-Posed Problems*; Wiley: New York, NY, USA, 1977.

28. Isakov, V. *Inverse Sourse Problems*; American Mathematical Society: Providence, RI, USA, 1996; pp. 1–14.

29. Licns, J.L.; Magenes, E. *Non-Homogeneous Boundary Value Problems and Applications*; Springer: Berlin, Germany, 1972; Volume 1, pp. 109–226

30. Krein, S.G. *Linear Differential Equations in Banach Space*; Transl. Math. Monogr.; American Mathematical Society: Providence, RI, USA, 1971; pp. 3–105.

31. Zalesny, V.B.; Diansky, N.A.; Fomin, V.V.; Moshonkin, S.N.; Demyshev, S.G. Numerical model of the circulation of the Black Sea and the Sea of Azov. *Russ. J. Numer. Anal. Math. Model.* **2012**, *27*, 95–111. [CrossRef]

Journal of
*Marine Science
and Engineering*

Article

Parallel Implementation of a PETSc-Based Framework for the General Curvilinear Coastal Ocean Model

Manuel Valera [1,*], **Mary P. Thomas** [1,2] and **Mariangel Garcia** [1,3] and **Jose E. Castillo** [1]

1   Computational Science Research Center, San Diego State University, San Diego, CA 92182-1245, USA;
     mthomas@sdsc.edu (M.P.T.); mariangel.garcia@tecnico.ulisboa.pt (M.G.); jcastillo@sdsu.edu (J.E.C.)
2   San Diego Supercomputer Center, University of California San Diego, La Jolla, CA 92093-0505, USA
3   Marine Environment and Technology Center (MARETEC), 1049-001 Lisbon, Portugal
*   Correspondence: mvalera-w@sdsu.edu; Tel.: +1-619-724-3073

Received: 24 April 2019; Accepted: 10 June 2019; Published: 13 June 2019

Abstract: The General Curvilinear Coastal Ocean Model (GCCOM) is a 3D curvilinear, structured-mesh, non-hydrostatic, large-eddy simulation model that is capable of running oceanic simulations. GCCOM is an inherently computationally expensive model: it uses an elliptic solver for the dynamic pressure; meter-scale simulations requiring memory footprints on the order of 10^{12} cells and terabytes of output data. As a solution for parallel optimization, the Fortran-interfaced Portable–Extensible Toolkit for Scientific Computation (PETSc) library was chosen as a framework to help reduce the complexity of managing the 3D geometry, to improve parallel algorithm design, and to provide a parallelized linear system solver and preconditioner. GCCOM discretizations are based on an Arakawa-C staggered grid, and PETSc DMDA (Data Management for Distributed Arrays) objects were used to provide communication and domain ownership management of the resultant multi-dimensional arrays, while the fully curvilinear Laplacian system for pressure is solved by the PETSc linear solver routines. In this paper, the framework design and architecture are described in detail, and results are presented that demonstrate the multiscale capabilities of the model and the parallel framework to 240 cores over domains of order 10^7 total cells per variable, and the correctness and performance of the multiphysics aspects of the model for a baseline experiment stratified seamount.

Keywords: high performance computing; HPC; PETSc; parallelization; scalability; parallel performance; streams; curvilinear; non-hydrostatic; ocean modeling; GCCOM

---

1. Introduction

As computational modeling and its resources becomes ubiquitous, numerical solutions to complex equations can be solved with increasing resolution and accuracy. At the same time, as more variables and processes are taken into account, and spatial and temporal resolutions are increased to model real field-scale events, models become more complex yet resource efficiency remains an important requirement. Multiscale and multiphysics modeling encompasses these factors and relies on High-Performance Computing (HPC) resources and services to solve problems effectively [1]. The models used for atmospheric and ocean studies are examples of such applications.

In atmospheric and ocean studies, one of the major challenges is the simulation of coastal ocean dynamics due to the vast range of length and time scales required: tidal processes and oceanic currents happen in fractions of days; wavelengths are scale lengths of kilometers; mixing and turbulence events need to be resolved at the meter or submeter scale; and time resolution is in terms of years to minutes or seconds. Consequently, field scale simulations need to cover both sides of the time/space spectrum while preserving accuracy, forcing the use of highly detailed grids of considerable size. The General Curvilinear Coastal Ocean Model (GCCOM) is a 3D, curvilinear, large-eddy simulation model designed specifically for high-resolution (meter scale) simulations of coastal regions, capable of capturing nonhydrostatic flow dynamics [2].

Another major challenge is the need to solve for nonhydrostatic pressures. Ocean and climate models are generally hydrostatic, limiting the physics they are able to capture, and large scale, so that they are computationally efficient enough to make forecasts in reasonable run-times. Historically, these models have opted to solve the more computationally efficient, but less accurate, hydrostatic version of the Navier–Stokes equations because simulating nonhydrostatic processes using the Boussinesq approximation is computationally intensive, as these models require solving an elliptic problem [3–5]. However, advances in HPC systems and algorithms have enabled the use of nonhydrostatic solvers in ocean models including GCCOM [6], MITgcm [7], SUNTANS [8], FVCOM [9], SOMAR [10], and KLOCAW [11].

A differentiating feature of these models is the method used to solve the nonhydrostatic pressure: the pressure can be either solved in the physical grid after taking the divergence of the momentum equation (Boussinesq), or reconstructed from the equation of state when density is chosen as the main scalar argument. In GCCOM, the pressure is solved on a fully 3D (not 2D plus sigma coordinates) computational grid (a normalized representation of the physical grid) that is created after applying a unique, general 3D curvilinear transformation to the Laplacian problem [2]. All computations are performed on the computational grid, which increases the accuracy of the results. The curvilinear coordinates formulation enforces the computational grid to be unitary and orthogonal, thus simplifying the boundary treatment and ensuring minimal energy loss by transforming any boundary geometry to an unitary cube, which in practice reduces the application of boundaries to a plane (or edge) of this cube. In contrast, the typical approach of approximating the grid cell to the problem geometry loses energy in the boundary proportional to the grid resolution. A detailed comparison of the GCCOM model with other non hydrostatic ocean models functionalities can be found in [12].

One other defining characteristic of some of these nonhydrostatic models is the HPC framework used: many use the Message Passing Interfae (MPI) framework (in some cases enhanced by using OpenMP, or accelerated using GPU resources), parallel file IO, and have large teams developing advanced multiphysics modules. Typically, these models are often nested within large, global, weather prediction models such as those used in disaster response situations, which is a long-term goal of this project. GCCOM is a modern model: written in Fortran 95; modular; employing NetCDF to manage the data. The first parallel version of an earlier GCCOM model (UCOAM) used a customized MPI-based parallel framework that scaled to a few hundred cores [13]. The UCOAM model has also been used for nesting inside the California ROMS global ocean model [14,15]. Having demonstrated the multiscale and multiphysics capabilities of GCCOM [12], the next phase of development requires a more advanced HPC framework in order to speed up development and testing and allow us to work with more realistic domains and algorithms. These factors motivated the decision to migrate the model to a more advanced HPC framework.

One additional factor in our project is the size of the team, and the level of effort needed to produce and support this type of model. Often the larger model projects are optimized for specific hardware that may not be portable. These models often require large development teams. One of the goals of the GCCOM project is to deliver a portable model that can run in a heterogeneous computing environment and proves useful to smaller research teams, but would someday scale to run inside of larger models. As the GCCOM model evolved, the PETSc (Portable-Extensible Toolkit for Scientific Computing) model was chosen to improve the scaling and efficiency of the model and to improve access to advanced mathematical libraries and tools [16–18]. The results presented in the paper justify the time and effort required to accomplish these objectives.

In this paper, the outcome of this approach is presented. The background of the model, motivation for parallelization, the parallel approach, and the choices for the underlying software stack are described in Sections 2.1, 2.2, and 2.4. The methodology and specifics of the test case used to validate the parallel framework are described in Section 2.5 and the validated results showing that the parallelized model produces correct results are presented in Section 3.1. The parallel performance is presented in Section 3.2 with results showing that the prototype model scales with the PETSc framework and to the maximum size of the HPC system used for these tests. Section 3.3 contains results demonstrating the multiscale and multiphysics capabilities of the model. Conclusions and future work are discussed in Section 4.

2. Materials and Methods

2.1. The General Curvilinear Coastal Ocean Model (GCCOM)

The General Curvilinear Coastal Ocean Model (GCCOM) is a coastal ocean model that solves the three-dimensional Navier–Stokes equation with the Boussinesq approximation, a Large Eddy Simulation (LES) formulation is implemented with a subgrid-scale model, capable of handling strongly stratified environments [12]. GCCOM features include: an embedded fully 3D curvilinear transformation, which makes it uniquely equipped to handle non-convex features in every direction including along the vertical axis [2]; a full 3D curvilinear Laplacian operator that solve a 3D nonhydrostatic pressure equation that accurately reproduces features resulting from the interaction of currents and steep bathymetries; and ability to calculate solutions from the sub-meter to the kilometer ranges in one simulation. With these key features, GCCOM has been used to simulate coastal ocean processes with multiscale simulations ranging from oceanic currents and internal waves [15] to turbulence mixing and bores formation, all in a single scenario [19], as well as the use of data assimilation capabilities [20], including thermodynamics and turbulence mixing on high-resolution grids, up to meter and sub-meter scales. Additionally, a key goal of the GCCOM model is to study turbulent mixing and internal waves in the coastal region, as a way to bridge the work of regional models (e.g., ROMS [21] or POP [22]), and global models (e.g., MPAS [23], HYCOM [24]), all the way to the sea–land interface.

GCCOM was recently validated for stratified oceanographic processes in [12] and has been coupled with ROMS [15] using nested grids to obtain greater resolution in a region of interest, using approaches that are similar to the work of other coupled ocean model systems. In [25], an overset grid method is used to couple a hydrostatic, large scale, coastal ocean flow model (FVCOM or unstructured grid finite volume coastal ocean model) with a non-hydrostatic model tailored for high-fidelity, unsteady, small scale flows (SIFOM or solver for incompressible flow on overset meshes). This method presents a way to offset the computational cost of a single, comprehensive model capable of dealing with multiple types of physics. GCCOM is similar to SIFOM in the curvilinear transformation and multiphysics capabilities, but they differ in the basic grid layout (staggered vs. non-staggered grids) and in the numerical methods each one applies to solve the Navier–Stokes equations. In addition, GCCOM has been taking strides in its development so it does not need to couple with a large-scale model to capture multiscale processes. Instead, GCCOM includes the entire domain in the curvilinear, nonhydrostatic, multiphysics capable region, where we handle thermodynamics, hydrostatic and nonhydrostatic pressure and equation of state density distributions at the same time. In this sense, GCCOM is a model capable of supporting both multiphysics and multiscale calculations at high computational costs. Recently, the PETSc-based GCCOM model has been coupled with SWASH [26] to simulate surface waves and overcome the limitations of the rigid lid [27]. SWASH is a numerical tool used to simulate free-surface, rotational flow and transport phenomena in coastal waters, we used the hydrostatic component to capture and validate surface wave heights in a seamount testcase.

GCCOM development began with [28] introducing the 3D curvilinear coordinates transformation, along with practical applications including the Alarcon Seamount and Lake Valencia waters [29,30]. Later, the model was revised by [2] to add thermodynamic processes, the UNESCO Equation of State, and use of a simple Successive Over–Relaxation (SOR) algorithm to solve the non-hydrostatic pressure [31] on an Arakawa-C grid [32], and an upgrade of the model to Fortran 95. In order to accelerate convergence, the authors in [33] implemented an elliptic equation solver for pressure that used the Aggregation-based Algebraic Multigrid Library (AGMG) [34]. Additionally, an MPI-based model with the SOR solver was also developed around the same time by [35]. All of these development efforts demonstrated improvements in model accuracy and speedup, but there were still limitations running GCCOM because of the amount of time spent in the pressure solver.

After studying options available for parallelizing the pressure solver, the Portable Extensible Toolkit for Scientific Computing, PETSc [16–18] was chosen because of its ability to handle the complexities of the Arakawa-C staggered grid, its large collection of iterative Krylov subspace methods, and its ability to interface with other similar libraries. An additional benefit is that PETSc has been selected be part of the DOE Exascale computing project [36], which will allow our model to scale to very large coastal regions at high resolutions, and model developers can expect that the PETSc libraries will have long-term support.

A prototype PETSc hybrid model was developed by [37], in which the pressure solver was parallelized and inserted into the existing model. As expected, the implementation had performance limitations: the rest of the model was solved in serial and was not PETSc based, which has documented performance issues; and the Laplacian system vectors needed to be scattered at each iteration, solved, and then gathered back to the main processor where the rest of routines were runing. The approach, although not optimal, presented promising results: the computational time of the floating point operations performed inside the pressure solver routine scaled by the number of processors on a single node. Total run-time was dominated by the time required to transfer data between processors before and after the floating operations. A scatter and gather call at each iteration drove up the communication time, and the decrease in computation time inside the pressure solver routine did not offset it. To address these issues, a full parallelization strategy was developed by [38] through the use of PETSc DM/DMDA (Data Management for Distributed Arrays) objects [18]. DM/DMDA objects are domain decomposition tools that distribute arbitrary 3D meshes among processors. By using proper domain decomposition, and linear system parallel solving, full PETSc-based parallelization of the model has been completed (see Section 2.4).

2.2. The PETSc Libraries

PETSc is a modular set of libraries, data structures, and routines developed and maintained by Argonne National Laboratory and a thriving community around the world. Designed to solve scientific applications that are modeled by partial differential equations, PETSc has been shown to be particularly useful for computational fluid dynamics (CFD) problems. PETSc libraries support a variety of different numerical formulations, including finite element methods [39], finite volume [40] or finite difference as is the case for the GCCOM model. The wide array of fundamental tools for scientific computing, as linear and nonlinear, solvers and the parallel domain distribution and input/output protocols native to it, make it one of the strongest choices to port a proven code into a parallel framework. PETSc makes also possible to use GPUs, threads and MPI parallelism in a same model for different effects, further optimizing code performance. As of today, PETSc is supported by most of the XSEDE machines many large-scale NSF and DOE HPC systems in the US and it has become a fundamental tool in the scientific computing community.

The PETSc framework is designed to be used by large scale scientific applications. In fact, it is one of the software components that is on the list of the Department of Energy's Exascale Computing Roadmap [36]. Part of the exascale initiative is to develop "composable" software tools, where different PDE-based models can be directly coupled. PETSc will be developing libraries that will couple multiphysics models at scale. When properly implemented, PETSc applications will be capable of running multilevel, multidomain, multirate, and multiphysics algorithms for very large domains (billions of cells) [41,42]. Our expectation is that, by using PETSc for the parallel data distribution model and MPI communications, the parGCCOM model will be capable of scaling to very large numbers of cores and problem sizes and to support a wide variety of physics models via the PETSc libraries. In addition, PETSc supports OpenMP and GPU acceleration, which will be useful for application optimization.

In this paper, the strategies used to parallelize the GCCOM model using the PETSc libraries are presented, with the result that the model attains a reasonable speed up and scale in MPI systems up to 240 processors so far (the max number of processors on the system where these tests were run), while demonstrating preservation of the solution by testing with baseline experiments such as stratified seamount as seen in Section 3.1. In this manner, we want to emphasize the effort saved by using an established HPC toolkit, while still preserving the unique features of our model.

2.3. PETSc Development in GCCOM

A prototype PETSc-GCCOM hybrid model was developed by [37], in which the pressure solver was parallelized and inserted into the existing model. As expected, the implementation had performance limitations, since the rest of the model was solved in serial, and the Laplacian system vectors needed to be scattered at each iteration, solved, and then gathered back to the main processor where the rest of routines were running. The approach, although not optimal, presented promising results: the computational time of the floating point operations scaled by the number of processors inside the pressure solver routine. The total run-time was dominated by the time required to transfer data between processors before and after the floating operations. A scatter and gather call at each iteration drove up the communication time, and any decreases in the computation time inside the pressure solver routine did not offset the gather-scatter increases.

To address these issues, a full parallelization strategy was developed by [38] that utilizes the PETSc DM/DMDA (Data Management for Distributed Arrays) objects [18,43]. Data Management (DM) objects are used to manage communication between the algebraic structures in PETSc (Vec and Mat) and mesh data structures in PDE-based (or other) simulations. PETSc uses DMs to provide a simple way to employ parallelism via domain decomposition using objects known as Data Management for Distributed Arrays (DMDA objects [43]. DMDAs control parallel data layout and communication information including: the portion of data belonging to each processor (local domain); the location of neighboring processors and their own domains; and management of the parallel communication between domains [16].

In the GCCOM model, DMDA objects are used for all domain data decomposition and linear system parallel solving. This approach towards parallelization of the model is described below (in Section 2.4).

2.4. Model Parallelization

In this section, we describe the key aspects of the model that impact the parallelization of the model. This includes the use of the DMDA objects to manage data decomposition and message passing, the location of the scalars and velocities on the Arakawa-C grid, the hydrostatic pressure-gradient force (HPGF) and its impact on the data decomposition, the pressure calculations which dominate the computations.

GCCOM uses finite difference approximations to solve differential equations on curvilinear grids and are operated on via stencils. These grids need to be distributed across processors, updated independently, and require special cases to handle the data communication where the local domain ends. This set of operations are referred to as domain decomposition. Throughout the GCCOM code, specific DMDAs are created to manage the layout of multidimensional arrays for the velocities (u,v,w) and the temperature, salinity, pressure, and density (T,S,P,ρ) scalars. Each of the velocity components and the scalars are located at positions in the staggered grid that require special treatment in order to be distributed and updated correctly, and is described in more detail in Section 2.4.1.

Another key PETSc component that is utilized in the GCCOM model is the linear system solver. PETSc provides a way to apply iterative solvers for linear systems using MPI. At the same time, the most computationally intensive part of the GCCOM model involves solving a fully elliptic pressure-Poisson system. This approach is unique among most CFD models because it embeds the fully 3D curvilinear transformation in the solver. Details of this Laplacian are discussed in Section 2.4.4. The strength of using the PETSc libraries to solve linear systems lies in the ability to experiment with dozens of different iterative solvers and preconditioners. Another advantage of using PETSc is being able to use PETSc with well known external packages such as Trilinos, HYPRE and OpenCL [44–46] with minimal changes.

These key PETSc elements, domain decomposition and linear system solvers, provide the foundations needed to develop the core parallel framework of the GCCOM model, and help to keep development effort to a minimum. This proved to be very helpful for the small development team involved in this effort.

2.4.1. The Arakawa-C Grid

GCCOM defines the locations of its vectors and scalars on an Arakawa-C grid, where the components of flow are staggered in space [32]. In the C-grid, the u-velocity component of velocity is located at the west and east edges of the cell, v-velocity is located at the north and south edges, and pressure and other scalars are evaluated at cell centers (see Figure 1). Similarly, the w component will be in the front face of the 3D cube, going inwards. This staggered grid arrangement becomes the first challenge of the parallel overhaul because, for the DMDA objects, every component is regarded as co-located and are referenced by their lower-left grid point.

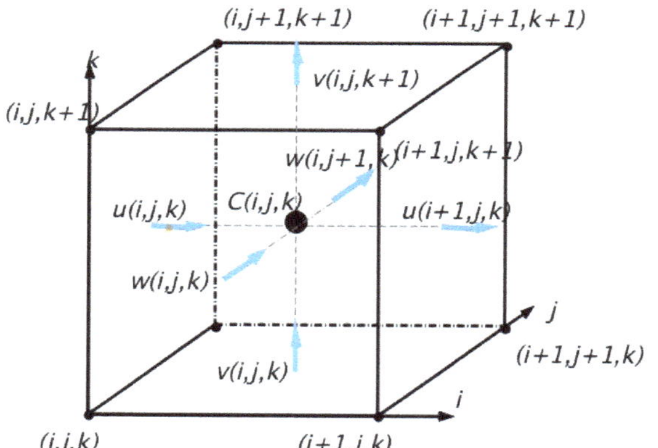

Figure 1. Diagram of the Arakawa C-grid, in which the velocity components u, v, and w are staggered by half a grid spacing. Scalars such as salinity S or temperature are located in the center.

The grid staggering results in different array *layouts* which depend on each variable position (u, v, w, p), which in turn creates different sizes for each layout. For example, there is one more u-velocity point in the horizontal direction than the v-velocity. Similarly, there is an extra v-velocity in the vertical direction compared to the u-velocity. Even though the computational ranges differ for each variable in the serial model, the arrays used by the parallel model were padded so that they would be the same dimensions. Note that this method is also used in the Regional Ocean Modeling System (ROMS) [47]. Table 1 describes the computational ranges and global sizes of each variable. The variables gx, gy, gz correspond to the full grid size in x,y,z and the fact the interior range is not the full array size counts for the number of ghost rows in each direction (e.g., $gx - 2$ has two ghost rows in that direction).

Table 1. Sizes and computational ranges of variable layouts on domain $\Omega = [gx \times gy \times gz]$ where gx, gy, gz are the total grid dimensions in each direction.

Variable	Size	Interior Range
Velocity u	$(gx) \times (gy - 1) \times (gz - 1)$	$0 : gx - 1, 0 : gy - 2, 0 : gz - 2$
Velocity v	$(gx - 1) \times (gy) \times (gz - 1)$	$0 : gx - 2, 0 : gy - 1, 0 : gz - 2$
Velocity w	$(gx - 1) \times (gy - 1) \times (gz)$	$0 : gx - 2, 0 : gy - 2, 0 : gz - 1$
Pressure	$(gx - 1) \times (gy - 1) \times (gz - 1)$	$0 : gx - 2, 0 : gy - 2, 0 : gz - 2$
Temperature	$(gx - 1) \times (gy - 1) \times (gz - 1)$	$0 : gx - 2, 0 : gy - 2, 0 : gz - 2$
Density	$(gx - 1) \times (gy - 1) \times (gz - 1)$	$0 : gx - 2, 0 : gy - 2, 0 : gz - 2$

One of the main features of the GCCOM model is the embedded fully 3D curvilinear transformation, capable of handling any kind of structured grid (i.e., rectangular, sigma, curvilinear) to be solved in the computational domain by these transformation metrics (a partial discussion of the transformation metrics is discussed in Section 2.4.4). The downside of this approach is a significant increase in memory allocation, since full sized arrays need to be stored for each of the transformation metrics. In GCCOM, the metrics arrays are associated with the location of the aforementioned variable layouts. This provides an opportunity to minimize overhead: the model arrays are grouped by variable layouts and similar functions inside the code. Similarly, other GCCOM modules such as Sub–Grid Scale (SGS) calculations and stratification (thermodynamics) handling, follow the same parallelization treatment. Table 2 shows the total number of DM objects (layouts) used by GCCOM and the associated variables.

Table 2. GCCOM Data Managment Objects (DMs) and variable layout used.

Name	Number of DMs	Variable Layout
daGrid	3	p
daSingle	1	p
daConstants	6	p
daSgs	1	p
daLaplacian	3	p
daMetrics	3	u,v,w
daCenters	3	p
daPressureCoeffs	10	p
daDivSgs	3	u,v,w
daDummy	1	p
daVelocities	3	u,v,w
daScalars	1	p
daDensity	1	p

Each of these DMs control the domain distribution independently, and each spawn the full sized arrays that become the grid where the finite difference calculations are carried out. In total, more than 100 hundred full sized arrays are used inside the model on different occasions, but around 60 of them remain in active memory because they are part of the core calculations, and all of them are distributed thanks to the use of these structures.

2.4.2. Domain Decomposition

Domain decomposition refers to dividing a large domain into smaller subdomains so that it can be solved independently on each processor. However, Grid operations with carried dependencies from one grid point to the next (self recurrence) are difficult to parallelize on distributed grids, thus truncating parallel implementation in this direction. In GCCOM, there are two such reasons that prevent domain decomposition in arbitrary directions: grid size and the calculation of the hydrostatic pressure-gradient force (HPGF) algorithm. For some of the experiments in GCCOM, the number of points in the y-direction is the minimum 6, used to simulate 2D processes, which in turn makes it too small to partition. The case of the parallelization of the HPGF is discussed next.

2.4.3. Hydrostatic Pressure-Gradient Force

The hydrostatic pressure-gradient force (Equation (1)) is calculated as a spline reconstruction and integral along the vertical (z-direction), which is updated at every time step as a fundamental step in the main GCCOM algorithm [12]. The requirement for vertical integration enforces a recursive computation inside the loop, i.e., subsequent values depend on previously calculated vertical levels, as can be seen in Equation (5) in which h is dependent of the whole column above itself. This set of Equations (1)–(5) describe a spline reconstruction of the hydrostatic pressure p_H over the z-column using a fourth-order approximation for $f_k(\xi) = \partial \rho(\xi)/\partial x$, by a series of coefficients (Δ_k, d_k) defined by the local change in vertical coordinate $h = z_{k+1} - z_k$:

$$\frac{\partial p_H}{\partial x} = \frac{\partial}{\partial x} \int_z^0 g\rho \, d\tilde{z}, \tag{1}$$

$$f(\xi) = f^{(0)} + f^{(1)}\xi + f^{(2)}\frac{\xi^2}{2} + f^{(3)}\frac{\xi^3}{6}, \tag{2}$$

$$f^{(0)} = f_k, \qquad\qquad\qquad f^{(1)} = d_k, \tag{3}$$

$$f^{(2)} = \frac{6\Delta_k - 2d_k - 4d_k}{h}, \qquad f^{(3)} = \frac{6d_k + 2d_{k+1} - 12\Delta_k}{h^2}, \tag{4}$$

$$h = z_{k+1} - z_k, \qquad \Delta_k = \frac{f_{k+1} - f_k}{h}, \qquad d_k = \frac{2\Delta_k\Delta_{k-1}}{\Delta_k + \Delta_{k-1}}. \tag{5}$$

As seen, the self-recurrence lies inside the algorithm and thus cannot be automatically detected by PETSc, resulting in a crash. The solution applied is not to partition data in the z-direction, which is easily done in PETSc by the command line `-da_processors_z 1`, another strength of the library. The result is that each processor stores and calculates the pressure-gradients on their respective sub-domain, effectively a vertical column. While this strategy enables the use of the HPGF, it also limits the parallelism available to solve the equations, a limitation that we will need to overcome in future developments of the model.

2.4.4. Laplacian Transformation

GCCOM solves the full nonhydrostatic 3D Navier–Stokes equations as follows:

$$\frac{\partial \vec{u}}{\partial t} + \vec{u} \cdot \nabla \vec{u} = -\frac{1}{\rho_0}\nabla p - \frac{g\rho}{\rho_0}\vec{k} - \nabla \cdot \vec{\tau}, \tag{6}$$

$$\frac{\partial T}{\partial t} + \vec{u} \cdot \nabla T = \nabla \cdot (k_T \nabla T), \tag{7}$$

$$\frac{\partial S}{\partial t} + \vec{u} \cdot \nabla S = \nabla \cdot (k_S \nabla S), \tag{8}$$

$$\nabla \cdot \vec{u} = 0, \tag{9}$$

$$\rho = \rho(T, S, p). \tag{10}$$

Here, $\vec{u} = (u, v, w)$ are velocities, $\frac{g\rho}{\rho_0}\vec{k}$ is gravity acceleration, $\vec{\tau}$ our stress tensor solved by Large Eddie Simulation, ρ is a equation of state and $k_{T,S}$ diffusivity constants for temperature and salinity.

The Boussinesq approximation [48] is applied to Equation (6) by taking the divergence and substituting $\nabla \vec{u}$ as in Equation (9). This step cancels every term other than the pressure p from

Equation (6), leaving a homogeneous Laplacian problem to solve as in Equation (11), which by itself is computationally expensive to solve, as we will discuss in the rest of this section:

$$\nabla^2 p = \frac{\partial^2 p}{\partial x} + \frac{\partial^2 p}{\partial y} + \frac{\partial^2 p}{\partial z} = 0, \tag{11}$$

$$
\begin{aligned}
\nabla^2 p = L(p) - L(x)\left[\xi_x \frac{\partial p}{\partial \xi} + \eta_x \frac{\partial p}{\partial \eta} + \zeta_x \frac{\partial p}{\partial \zeta}\right] \\
- L(y)\left[\xi_y \frac{\partial p}{\partial \xi} + \eta_y \frac{\partial p}{\partial \eta} + \zeta_y \frac{\partial p}{\partial \zeta}\right] \\
- L(z)\left[\xi_z \frac{\partial p}{\partial \xi} + \eta_z \frac{\partial p}{\partial \eta} + \zeta_z \frac{\partial p}{\partial \zeta}\right].
\end{aligned}
\tag{12}
$$

The GCCOM Laplacian in curvilinear coordinates is formulated in [29]; note that we are following notation from [31], the operator $L()$ is defined in Equation (13), where a, b, c, d, e, q are coefficients related to the curvilinear transformation [12] that are defined in terms of the Jacobian J and generalized in two sets of coordinates by their even commutation: $\{x, y, z\}$ as $\{1, 2, 3\}$, and $\{\xi, \eta, \zeta\}$ as $\{A, B, C\}$, a general rule for the derivatives can be defined as in Equation (14), where we have adopted a compact differential representation, i.e., $A_1 = \frac{\partial \xi}{\partial x}$,

$$L() = a\frac{\partial^2()}{\partial \xi^2} + b\frac{\partial^2()}{\partial \eta^2} + c\frac{\partial^2()}{\partial \zeta^2} + 2\left[d\frac{\partial^2()}{\partial \xi \partial \eta} + e\frac{\partial^2()}{\partial \zeta \partial \eta} + q\frac{\partial^2()}{\partial \xi \partial \zeta}\right], \tag{13}$$

$$A_1 = \frac{\partial \xi}{\partial x} = J(2_B 3_C - 2_C 3_B). \tag{14}$$

By applying these elements to the transformed Laplacian operator and discretizing using 2nd-order finite difference, Equation (15) becomes the discretized transformed Laplacian operator in general curvilinear coordinates [33],

$$
\begin{aligned}
\nabla^2 p = -\frac{1}{2(\Delta_\xi^2 \Delta_\eta^2 \Delta_\zeta^2)}\Big\{ & 4\alpha(i,j,k)p(i,j,k) \\
& + [\beta_1(i,j,k) + \beta_2(i,j,k)]p(i+1,j,k) \\
& + [\beta_1(i,j,k) - \beta_2(i,j,k)]p(i-1,j,k) \\
& + [\lambda_1(i,j,k) + \lambda_2(i,j,k)]p(i,j+1,k) \\
& + [\lambda_1(i,j,k) - \lambda_2(i,j,k)]p(i,j-1,k) \\
& + [\tau_1(i,j,k) + \tau_2(i,j,k)]p(i,j,k+1) \\
& + [\tau_1(i,j,k) - \tau_2(i,j,k)]p(i,j,k-1) \\
& - \tau_{xy}(i,j,k)(p(i+1,j+1,k) + p(i-1,j-1,k)) \\
& + \tau_{xy}(i,j,k)(p(i+1,j-1,k) + p(i-1,j+1,k)) \\
& - \tau_{yz}(i,j,k)(p(i,j-1,k-1) + p(i,j+1,k+1)) \\
& + \tau_{yz}(i,j,k)(p(i,j-1,k+1) + p(i,j+1,k-1)) \\
& - \tau_{xz}(i,j,k)(p(i-1,j,k-1) + p(i+1,j,k+1)) \\
& + \tau_{xz}(i,j,k)(p(i-1,j,k+1) + p(i+1,j,k-1))\Big\},
\end{aligned}
\tag{15}
$$

where α, β_1, β_2, λ_1, λ_2, τ_1, τ_2, τ_{xy}, τ_{yz}, τ_{xz} are transformation coefficients found after algebraic manipulation.

Now, the pressure can be solved entirely as a system of linear equations in the form $Ax = b$, where A is the system matrix constructed by the position coefficients, b is the known pressure on each point (Right Hand Side, or RHS), and x is the solution vector for pressure.

A peculiarity of this system is the inherent lexicographical ordering: the points of the 19-point stencil follow a specific positioning pattern in the program memory, as seen in Figure 2: on each xy-plane, points are sorted in the y-direction before the x-direction, then the points are sorted by planes on the bottom z-axis first; this is the same as a front-to-back, bottom-up, ordering in the 3D stencil. This bookkeeping is crucial to obtain the right dynamics out of the Laplacian.

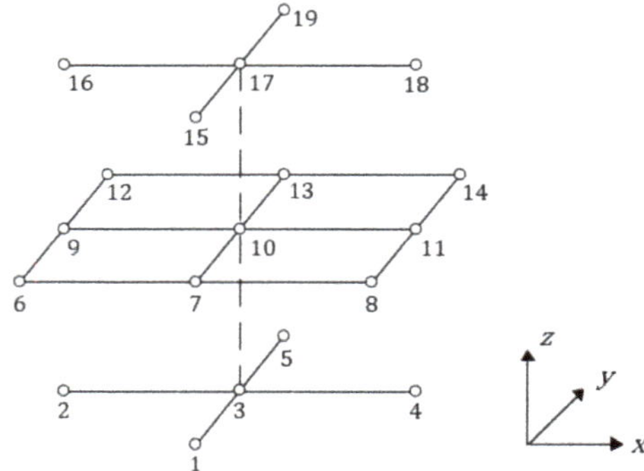

Figure 2. Lexicograpical ordering of the 19-point stencil. Here, element 10 is the current point of the stencil [38].

It is important to note that the coefficient matrix, A is large ($[gx, gy, gz] \times [gx, gy, gz]$), sparse, and in general not singular, and non-symmetric [33]. This automatically eliminates methods that directly invert the matrix to solve $x = A^{-1}b$ for large problem sizes. In GCCOM, this equation is solved with the Generalized Conjugate Residual (GCR) method preconditioned with block Jacobi. At this point, we are able to take advantage of the PETSc linear solver system which includes a long list of solvers and preconditioners that can be accessed via command line arguments: `-ksp_type [solver] -pc_type [precond]`.

2.4.5. External Boundary Data

External boundary conditions are generated outside of GCCOM and stored separately in ASCII files. They contain velocity data at the boundaries for planes in the x-direction (west and east) and z-direction (north and south). Processors located on these boundaries read boundary condition data into local memory. As we work with higher resolution meshes, this external file reading becomes an obstacle to overcome and adds to the I/O overhead. Part of the required future work includes updating these routines to read parallel NetCDF files.

2.5. Test Case Experiments

In this section, we describe the experiments used to validate the parallel framework of GCCOM. The Seamount experiment was chosen for being the most comprehensive for the model, using most—if not all—of the model capabilities at once.

2.5.1. Test System

Timings for test cases were primarily conducted on the Coastal Ocean Dynamics (COD) cluster at the Computational Science Research Center at San Diego State University, a Linux based cluster with the following features:

- 352 Intel *Xeon* Processor E5-2640 v4 (2.40 GHz) across 19 nodes,
- 7 nodes comprised of 16 processors each with 65 GB RAM per node,
- 12 nodes of 20 processors and 263 GB RAM per node,
- 40 GB/s Infiniband network interconnect,
- High Performance GPFS file system.

GCCOM is a memory intensive model: there are over 100 matrices that must be held in memory. For the "small" test case run in these experiments, 6×10^7 cells per array, the model requires on the order of 2×10^2 GB of storage plus the run-time overhead. Consequently, the test cases used for this research were run on the large-memory nodes of the Coastal Ocean Dynamics (COD) system. The CCD system hosts 12 large memory nodes with 263 GB per node and 20 cores per node, for a total of 240 available cores. The total memory is 20×263 GB $\approx 10^4$ GB.

The rest of the machine nodes were not able to run a problem of this size because of the memory requirements of the model. As mentioned before, the scope of the research reported in this paper was to validate the PETSC-based implementation of the model physics, and to defer optimizations to later research. Of future interest will be to profile the memory consumed by the framework. Most of the timing data was recorded using the built-in PETSc timers, which have been shown to be to be fairly accurate and to have little or no overhead [16].

In addition, the model has been ported to the XSEDE Comet machine at the San Diego Supercomputer Center, in preparation for running larger jobs and as a test to check the portability of the model [49]. Comet currently has 1944 nodes with 320 GB/node, four large memory nodes with 1.5 TB of DRAM and four Haswell processors with 16 cores per node. Future plans include exploring how the model performs on this type of system.

2.5.2. Stratified Seamount

Seamount experiments are regarded as a straightforward way to showcase an oceanographic models' capabilities and behavior. We carried out our timing and validation tests on a classical seamount, using continuous stratification with temperature ranging between (10 °C to 12 °C from the bottom to the top in the water column) and equivalent density for seawater, while maintaining the salinity constant at 35. The bathymetry for this experiment is defined by Equation (16),

$$D(x,y) = L(-1 + a * e^{-b(x^2+y^2)}),\qquad(16)$$

where $L = 1000$ m is the maximum depth and characteristic length, and the parameters $a = 0.5$ and $b = 8$ control the seamount shape. The experimental domain is $(x,y,z) = 3.6$ km $\times$ 2.8 km $\times$ 1 km. The experiment is forced externally with a linearly increasing $u-$velocity on the vertical column from 0 at the bottom to 0.01 m/s at the top, coming from the east direction.

The grid was created with cell clustering at the bottom, along half the domain in each horizontal direction as seen in Figure 3; this created a 3D curvilinear grid with the point distribution seen in Equation (18), where L_i is the dimension length, D is the horizontal clustering position ($D = L_i/0.5$) and β varies between {1,5} uniformly along the vertical, $\beta = 5$ at the bottom where the cell clustering is most and $\beta = 1$ at the surface where there is no clustering. This grid is based in the work of [2], expanding it to be able to use continuous stratification and simplifying the curvilinear implementation:

$$x_i = D\{1 + \frac{sinh[\beta(\xi_i - A)]}{sinh\beta A}\},\tag{17}$$

$$A = \frac{1}{2\beta}ln[\frac{1 + (e^\beta - 1)(D/L_i)}{1 + (e^{-\beta} - 1)(D/L_i)}].\tag{18}$$

Three grids were generated where each would have enough resolution to be able to show strong scaling on the test cluster (see Section 3.2). The grid sizes are $(x, y, z) = 1500 \times 100 \times 50$ for the smallest one, with 7.5 million cell points per variable (our lowest resolution problem), a second grid with sizes $(x, y, z) = 2000 \times 100 \times 100$ having 20 million points per variable, and a high resolution problem of size $(x, y, z) = 3000 \times 200 \times 100$ yielding around 60 million cell points per variable. The simulation was run for five main loop iterations which in turn is 5 s of simulation, creating and writing a NetCDF file as output. This I/O operation happens twice and is removed from the parallel timing and performance analysis since it is not yet parallelized.

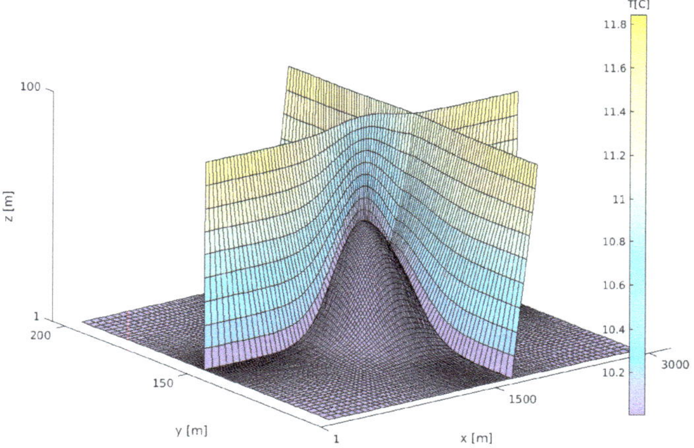

Figure 3. Seamount grid of size $x = 3000, y = 200, z = 100$. Point agglomeration can be seen along each half horizontal direction with parameter $\beta = 5$, gradually decreasing onto the top to be uniform ($\beta = 1$).

3. Results and Discussion

3.1. Model Validation

This section shows how the parallelized model produces correct results for our stratified seamount experiment. Correct in this context means replicating the same results obtained by the serial version of the model which was recently validated for non-hydrostatic oceanographic applications [12]. We also discuss the strategy applied to compare the models output and how much the results differ when adding more processors. Comparison is presented against the serial and parallel outputs for a single processor, to give a rounded picture on how communication errors are propagated in the parallel framework.

3.1.1. Validation Procedure

The Stratified seamount experiment is run for five computational iterations, or cycles of solving Equations (6)–(10) inside the main computational loop. The goal of this exercise is to define the consistency of the whole computational suite in the parallel framework, compared to the same set of equations being solved in serial. Each iteration represents a second of simulation time for a total of $5[s]$. This validation process is carried out in the high resolution problem ($3000 \times 200 \times 100$).

NCO operators [50] were used to obtain the root-mean squared (RMS) error of the output directly from the NetCDF output files, comparing each parallel run with the single processor serial run as seen in Table A1 and also with the single processor parallel run, visible in Table A2. The RMS is obtained for each main variable (p,u,v,w,T,D) by (1) subtracting each parallel output from the single processor output, (2) applying a weighted average with respect to time in order to unify the records obtained in a single time-averaged snapshot, and (3) obtaining the RMS error using the ncra operator. For the results, we report the maximum absolute value of the RMS, minimizing boundary errors by reading the 50th X-Y plane out of the vertical column of 100 planes.

3.1.2. Comparison with Serial GCCOM Model

Here, we present in table form the values of the maximum RMS along the half point of the vertical column for each of the parallel runs obtained while comparing with the serial output of the identical experiment of the Serial GCCOM. Note that we will refer to the comparison of parallel results to serial as *vs. serial* for the rest of the document.

As can be seen models are in agreement for every practical purpose, with exception of the pressure (which is the result of solving the linear system and depends of a krylov subspace solver method, and therefore can be refined) placing the biggest error at around 10^{-5}. The velocities readings are all of them between 10^{-7}–10^{-8} and the scalars D, T are close to machine error. Additionally, from Table A1, errors are virtually equal across all parallel runs. This tells us the solutions we obtain from the parallel model are in very close agreement with the serial model. This comparison brings confidence to the robustness of the PETSc implementation we have achieved, yielding the same degree of error beyond the data partitioning used. Nonetheless communication and rounding errors exist and the number of processors used are affected by them as we will explain next.

3.1.3. Error Propagation

In this section, we examine how solution errors grow along with number of processors/nodes when running the exact same experiment. Often, these errors are a consequence of halo communication and rounding errors. The results can be seen in Table A2 and Figure 4. Note that, for these tests, where we are comparing parallel model output for one processor vs. N processors, we will refer to this as *vs. parallel*. In every case, the *vs. serial* RMS error is below 10^{-5} and would be unable to influence the dynamics of our experiment, effectively transferring the physics model validation obtained at [12]. In addition, in the case of *vs. parallel*, RMS error is in every case orders of magnitude smaller than *vs. serial*; as this is the case, we can confidently conclude that using as many as 240 processors (and presumably more) won't affect the solution because of rounding or communication errors that may otherwise be introduced by a large data distribution layout. This finding brings confidence in our parallel-enabled model.

In this section, we have shown that the PETSc based parallel GCCOM framework preserves the solutions obtained by the validated serial GCCOM model for different mesh sizes of the Seamount test case. We have also shown here that the communication errors PETSc introduces are small enough not to be a problem with the 240 processors/12 nodes we have used.

Finally, the trend we show points that for the communication errors (*vs. parallel* error) to catch up with the parallel framework migration error (*vs. serial* error) we would need to double the processor count with a properly sized experiment, something that would be impractical to run in the serial model. In short, we have attained a new range of problem sizes we can solve in this new parallel framework, while carrying out the physics validation obtained in the serial version of the model.

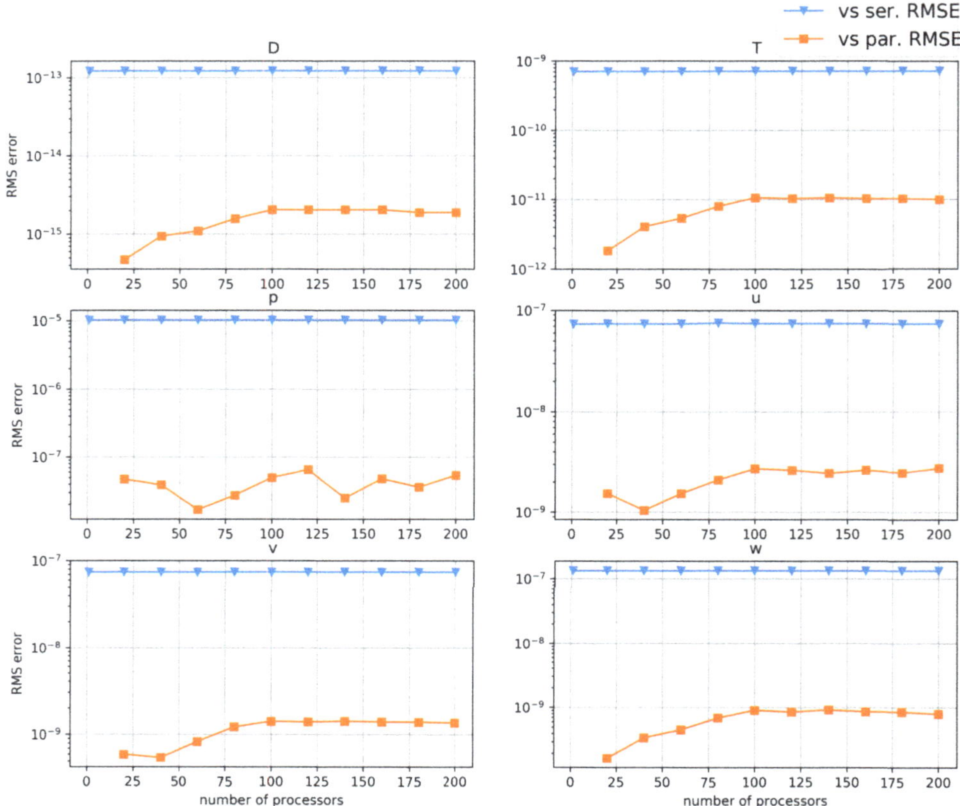

Figure 4. Error comparison between serial and parallel models for one processor output, the parallel error is consistently orders of magnitude smaller than the serial comparison, which in turn is small enough to carry the physics validation of the model. Comparison with Serial GCCOM Mode.

3.2. Model Performance

Model performance can be assessed and measured in different ways, but in general should improve with number of resources allocated, up to the limit of where the problem size and other factors make it limit or even degrade performance. This is known as scalability or scaling power.

In order to analyze the parallel performance of the PETSc-based GCCOM model, it is important to first validate the results as was done in Section 3.1. Once the model is validated, and the algorithmic approach has been verified, the next step is to identify critical blocks and bottlenecks in order to determine what elements of the model can be optimized. In the case of GCCOM, key factors include problem size and resolution, time step resolution, numerical methods and solvers, file IO, the PETSc framework, and the test cluster. The multiscale/multiphysics non-hydrostatic capabilities of the GCCOM model are demonstrated using the stratified seamount test case using different meshes and a 3D lock exchange test case. The impact and results of these factors are presented below.

3.2.1. PETSc Performance

The PETSc framework has its own performance characteristics, and basically defines an upper limit that we can expect from any model using the framework. The performance of PETSc is measured using its *streams* test, which outputs the speedup as a function of the number of cores [16]. Streams measures the communication overhead and efficiency that is realistically attainable in a system. The test probes the machine for its maximum memory speed, and becomes an alternative way to measure the maximum bandwidth, speedup and efficiency. The Streams tests are done as part of the PETSc installation process and is regarded as an upper limit of the speedup attainable on the system, limited by the memory bandwidth. More information can bee seen the PETSc user guide (see [16] Section 14—Hints for Performance and Tuning).

The speedup is defined as the ratio of the runtime of the serial model (T_1) to the time, T_n, taken by the parallel model as a function of the number of cores (n). The ideal speedup of an application would be a perfect scaling of the serial (or base) timing to the number of processors used, or $S_{ideal} = T_1/n$. The measured speedup of a model is defined as follows $S_N = T_1/T_n$.

The PETSc streams speedup is a diagnostics test for our system. As we will see in Section 3.2.3 it shows as a linear trend, which indicates that the bandwidth communication capacity grows linearly across the system, in this case up to 240 processors on 12 nodes. Note also that the PETSc framework shows no sign of turning over, indicating that it is capable of scaling to a much larger number of cores.

In practice, most distributed memory applications are bounded by the memory bandwidth allocation of the system, which is measured in PETSc by the streams test. For every test performed in our analysis, we have used the streams' speedup estimate as an upper bound on the speedup that can be obtained as a result of the memory bandwidth constraint. For the GCCOM model, we have determined that, for large enough problem sizes, this speedup threshold can be surpassed, effectively offsetting this performance limit by some margin.

3.2.2. Profiling the GCCOM Model

To profile the GCCOM model, we analyze the three phases that are typical of many parallel models: initialization, computation, and finalization. Initial wall-clock profiling timings show that the time spent in the finalization phase is less than 1% of the wall-clock time, independent of problem size and number of cores. Consequently, it will not be part of further analysis discussed in this section. Timings also show that approximately 15–35% of total wall-clock time is spent in the initialization phase, where the PETSc arrays and objects are initialized, memory is allocated, initialization data are loaded in from files, and the curvilinear metrics are derived. The remainder of the execution time is spent in what we refer to as the "main loop", in which a set of iterative solutions to the governing equations are computed after the startup phase. In general, as the number of cores increases, the percentage of time spent in the main loop goes down, while the time spent in the initialization phase increases.

An explanation for the impact on scaling due to the startup phase may have something to do with the strategy employed to initialize and allocate the arrays used in the serial model. First, the model uses serial NetCDF to read and write data, effectively loading external files onto one master node and then scattering the data across the system. Similarly, when writing output data, the whole array is gathered onto one node, and then results are saved serially to a NetCDF file. Thus, the model is both IO and memory bound. This is a well known issue, and moving to parallel IO libraries is an important next step for this model.

In order to quantify the roles that these processes play in the total run time, we measured the partitioning of the total wall clock time as a function of the number of processors for three key functional areas: the main loop, or computational time; the I/O time; and the MPI communication time. The results can be seen in Figure 5, which shows a stacked histogram view of the functional area timings for the $3000 \times 200 \times 100$ problem. The figure plots the percentage of time used by each component as a function of the number of cores. In this figure, we see three trends: the computational time (the bottom, or blue, group of datum) dominates the run-time for small number of cores, and appears to scale well; the MPI communication time increases with the number of cores, which is expected and is a function of the model and the PETSc framework. The figure also shows clearly that I/O is impacting the run-time, and increasing with the number of processors. This would explain why the model is not scaling well overall. Stacked plots for the lower resolution grids are presented in Figure 6, here the overtaking of communication and I/O times over computation time is evident. These problems are too small to take real advantage of the MPI framework over the 240 processors system and are capped by the I/O and memory bandwidth speeds.

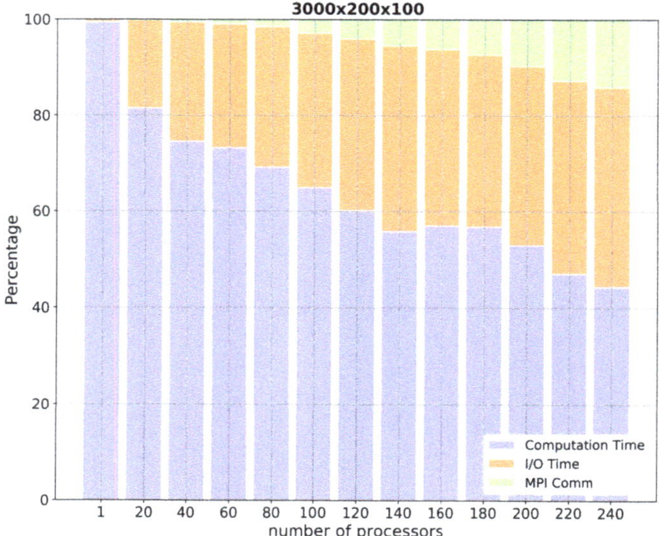

Figure 5. The histogram above shows a stacked normalized plot of the time partitioning between computation, I/O operations and estimated communication times as a function of processors for the high resolution problem.

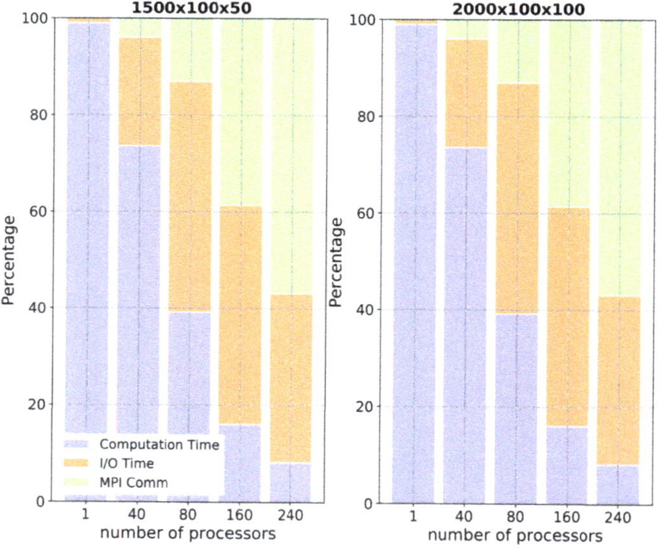

Figure 6. Stacked normalized plot of the time partitioning for the lower resolution grids.

As mentioned above, the primary goal of this paper is to report on advances made to the GCCOM model using the PETSc framework, to validate the results of the parallel version of the serial model (which is done in Section 3.1, and to show that the computational aspect of the model scales. Based on these goals, we will focus our scaling analysis primarily on the computational time required to solve the flow processes (knowing that we would be addressing parallel file I/O in future research). Thus, for the rest of this section, we will analyze the performance of the computational time, which is calculated as follows:

$$T_{comp} = T_{main\_loop} - T_{IO\_main\_loop}. \tag{19}$$

3.2.3. Parallel Performance Analysis

Ideal parallel performance is usually described as a reduction in execution time by a factor of the number of processors used. However, this is seldom achieved. Several factors can limit model speedup, some of which are discussed here, but a more generalized overview can be found in several well-known textbooks [51,52]. Examples include loop calculations that cannot be unrolled because the statements are dependent upon previous steps in the calculation, collecting information on all processors before computing the next step (a self-recurrent loop). In addition, the hardware of the system could potentially impact speed, including chip memory bandwidth or the network. Despite these factors, speedup can be achieved by splitting up the work between multiple processors, which reduces the calculation time.

Model performance can be assessed and measured in different ways, but in general should improve with number of resources allocated, up to the limit of where the problem size and other factors make it limit or even degrade performance. This is known as scalability or scaling power.

Figure 7 compares the speedup of the PETSc Streams test with the speedup of the GCCOM computational work done in the `main loop`, for the Seamount test cases, as function of the number of cores. The plot shows a linear trend, which is a consequence of the maximum memory bandwidth allocation, which increases with the growing number or processors. We can see that the bandwidth communication capacity grows linearly across the system, in this case up to 240 processors in 12 nodes. Interestingly, the high resolution seamount experiment speedup ($3000 \times 200 \times 100$) is consistently better than the streams test speedup. This is explained by the size of the high resolution problem benefiting from internal PETSc optimizations that occur within the DM and DMDA objects, which dynamically repartition grids and adjust the MPI communicators during the computations [43]. This is not the case for the lower resolution cases: for these, we see that the speedup trend follows the streams test closely, but it never surpasses the PETSc limit. In fact, we see the lower resolution trends being bogged down by too much data distribution and hence more message passing, and the speedup ends up being worse than the streams test with more processors added.

The measurement of the parallel efficiency indicates the percentage of efficiency for an application when increasing the number of available resources. Efficiency is defined by Equation (20):

$$E_n = \frac{T_1}{n * T_n}, \tag{20}$$

where T_1 and T_n are the execution times for one and n processors, respectively, and n is the number of processors. Ideal efficiency would be the case, for example, where doubling the number of processors halves the run time. Efficiency is expected to decrease when too many processors are allocated to a specific problem size. The optimal number of processors to use, from an economical perspective, can be determined from this metric. It is important to keep in mind that the application can show speedup while still decreasing its efficiency. The most common causes for decreasing efficiency are usually related to memory bandwidth speed limits and sub-optimal domain decomposition. As we will see later in this section, when these factors are taken into account, the scaling performance behavior can be explained.

Figure 8 shows efficiency for the GCCOM seamount test cases and the PETSc Streams test, calculated using Equation (20). From this, we can see that the PETSc Streams efficiency levels off at around 30%. This efficiency is typically regarded as the realistic efficiency of the system, limited by memory bandwidth, and as such we see that for the highest resolution problem ($3000 \times 200 \times 100$)

we are obtaining a better than streams efficiency across all nodes used, hinting at our parallelization overcoming the memory bandwidth overhead up to 240 processors. This efficiency is expected to decrease with a higher number of processors, something we don't yet see happening for the highest resolution case but does happen for the lower resolution experiments. Once again and as we saw with the speedup chart, these problems are still too small to take real advantage of the parallelization; although they present some speedup, the efficiency of resources allocated hardly justifies using more than a few nodes to run these problems.

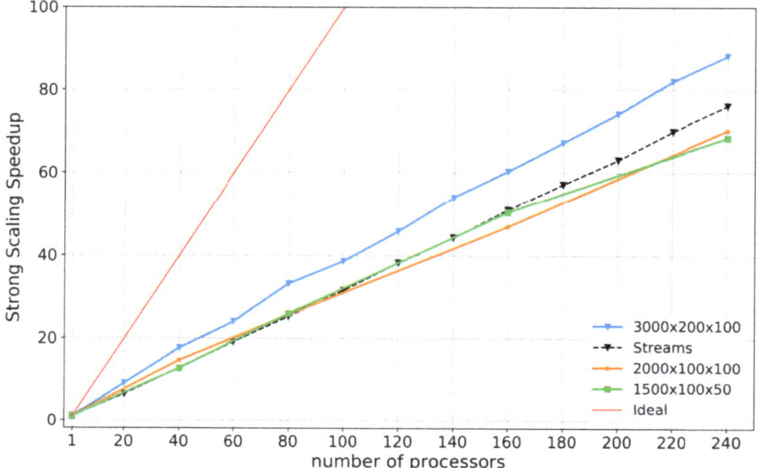

Figure 7. Speedup scaling for stratified seamount experiments of different resolutions compared to the Portable, Extensible Toolkit for Scientific Computation (PETSc) streams test results for the test system [16]. The theoretical ideal is shown for reference.

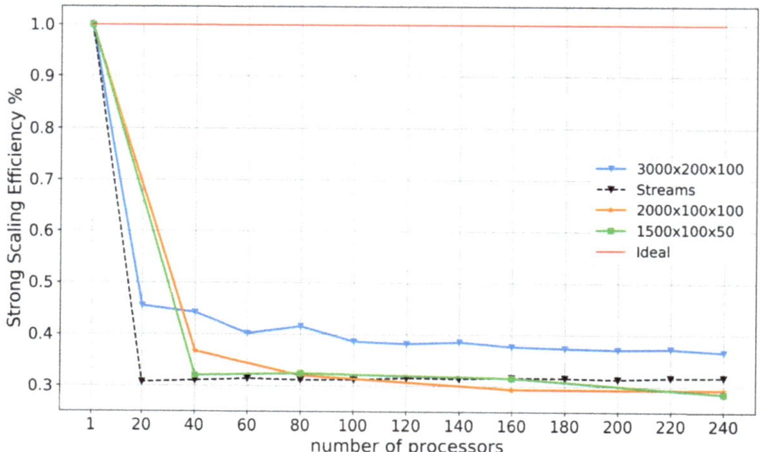

Figure 8. Efficiency for stratified seamount experiments of different resolutions compared to the PETSc streams test results for the test system [16]. A theoretical ideal is shown for reference.

Finally, given the restrictions imposed on the model in the form of self-recurrent algorithms and forced data distribution by HPGF algorithm, and keeping in mind that we have left out the I/O and initialization processes (since they have not been parallelized), we regard this version of parallel implementation to be a success. The model demonstrates efficiencies that are better than the streams test estimates for our stratified seamount experiment, yet preserving the solution to any practical threshold even when partitioning the problem across 200 processors and 12 nodes.

3.3. Multiscale/Multiphysics Capabilities

This section describes how GCCOM is capable of handling complicated fluid behavior by means of two different methods. First, the multiple physical processes simulated inside the model, such as hydrostatic and non-hydrostatic pressure, sub-grid scale turbulence, thermodynamics and the density equations of state for seawater work together to capture complex processes. Second, the grid resolution we use will have a major impact on the detail level and richness of the physics we obtain. In this section, we present results of comparing the output of two different resolutions in the Seamount case to better illustrate the point of the multiple physics and multiple scales GCCOM can capture.

Figures 9 and 10 show the velocity flow along the horizontal axis. The images compare and contrast the High (left) vs. Low (right) grid resolution details for both side and bottom views of the domain. The rows of images show a series of zoomed-in details, represented by the rectangular boxes. The bars to the right of each image depict the scale values.

A side view comparison between the seamount test case of $3000 \times 200 \times 100$ (high resolution, left side) and $1500 \times 100 \times 50$ grid points (low resolution, right side) is shown in Figure 9 for the horizontal velocity, being forced on the right side of this figure. The high resolution has twice the resolution in each direction and eight times the number of grid points. This snapshot was taken at the mid-section of the domain after $t = 6000$ s of simulation. Each problem has been run with the same conditions and shows similar behavior.

In the top row of images, the kilometer scale is depicted along the horizontal axis. The image shows an accumulation of contours at the base of the seamount, and somewhat uniform velocities over the rest of the domain for both of the resolutions. The difference between the two plots resides in the density and locations of the contour lines. For the high resolution, there are significantly more contour lines around the seamount bathymetry than for the low resolution plots. Row 2 of Figure 9 shows a zoomed in detail of these structures. Here, the images show marked differences between the high and low resolution cases. In this frame, we see that the features developed in the high resolution panel (left) are not captured in the low (right) resolution case. At the same time, the richness of the higher resolution case can be explored further, as is shown in the bottom left panel. The increased magnification reveals a series of eddies, while in the lower resolution counterpart no special behavior is seen. These results demonstrate that the GCCOM model is capable of capturing more information as the grid resolution increases: an increase in the number of points translates to capturing richer and more complex phenomena across the domain; and the multiscale processes, ranging from kilometer to meter scale lengths.

The ability of the GCCOM model to capture both high and low resolution flow features is seen once again in Figure 10, where the view is from the bottom plane of the domain, where the velocity flows from East to West. Again, the flow is captured at $t = 6000$ s. We can see structures developing widely in the high resolution problem, while the low resolution only shows them happening in specific spots and in a broader distribution, but we see no sign of high resolution fluid structures in the rest of the domain. The middle rows in Figure 10 show an important difference in contour details: while the low resolution grid shows a structure that is similar to that of the high resolution grid, the high resolution grid captures the meandering waves of low velocity fields on the bottom of the seamount, something we could see if we were modeling the shape of a sandy sea bottom. The details and number of eddies behind the seamount peak are also richer in the high resolution grid, while only one broad eddy-like structure is seen in the low resolution case (bottom panels). The results shown in this section demonstrate that GCCOM is capable of capturing different types of phenomena including fluid flow, nonhydrostatic pressure and thermodynamics, over scales that range from 10^0 to 10^3 m, thus establishing that GCCOM is both a multiphysics and a multiscale model.

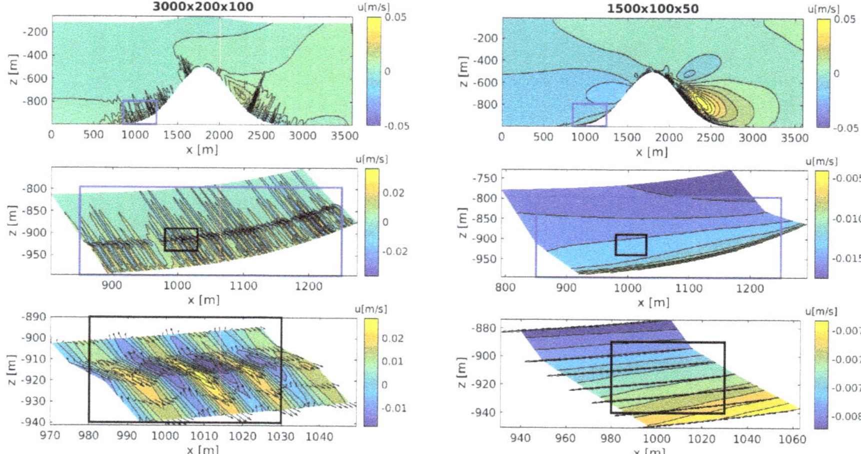

Figure 9. Figures depict zoomed-in details of the High (**left**) vs. Low (**right**) resolution side views of the stratified seamount experiment at $t = 6000$ s.

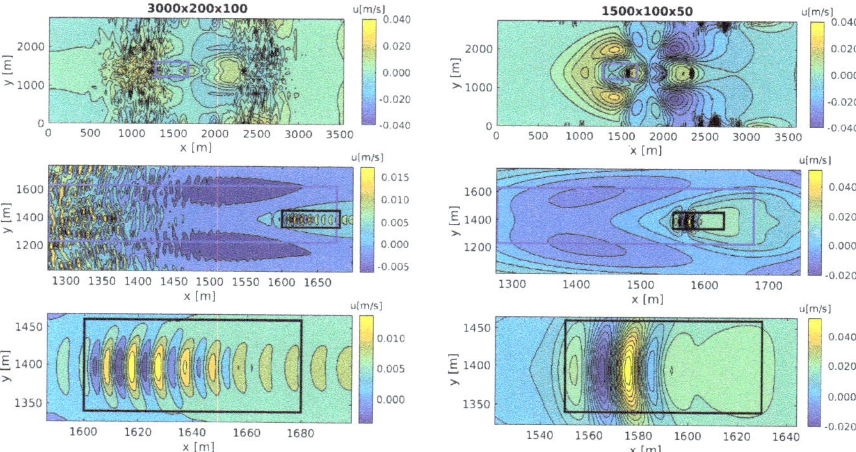

Figure 10. Figures depict zoomed-in details of the High (**left**) vs. Low (**right**) resolution bottom plane view for the stratified seamount experiment at $t = 6000$ s.

4. Conclusions

We have successfully implemented a parallel framework based on the serial GCCOM model, using domain decomposition and parallel linear solver methods from the PETSc libraries. The model has been validated using the seamount test case. Results show that the parallel version reproduces serial results to within acceptable ranges for key variables and scalars: around 10^{-5} for the Krylov subspace pressure solver; 10^{-7} to 10^{-8} for the velocities; and the scalars D and T are on the order of 32-bit machine precision.

Measured performance improvement tests show that detailed simulation run-times follow the scaling of the core PETSc framework speed tests (Streams). In some cases, the GCCOM model outperformed the streams test because the problem size is big enough to offset the communication overhead. For the experiments run in this study, the speedup was improved by a factor of 80 for 240 cores, and follows closely (or is better than) the speedup of the PETSc Streams test. Additionally, the gain in speedup shows that we can expect the model will be capable of additional improvement when it is migrated to a larger system with more memory and cores.

Utilization of the PETSc libraries has proven to be of significant benefit, but using PETSc has its pros and cons: We found that there was significant savings in the time needed for HPC model development, which has immense value for our small research group, but that the learning curve requires significant effort. For example, the complexity of representing the Arakawa-C staggered grid using Fortran Matrices and MPI communications schemes was extremely complex. This was more effectively achieved by employing the PETSc DM and DMDA parallelization paradigm for the array distribution and linear solvers. In addition, once completed, the model scaling improved, while adding and defining new scalars, variables, or testing different solvers was greatly simplified. We note that the development and testing of those objects was challenging, and eventually became the topic of a masters thesis project. Recently, PETSc has started offering staggered distributed arrays, DMSTAG (which represents a "staggered grid" or a structured cell complex), which is something we will explore in the future.

Based on our experiences, we strongly recommend PETSc as a proven alternative to obtain scalability in complex models without the need to build a custom parallel framework. As stated above, with PETSc, there is a learning curve: the migration of the model from an MPI based model to the current PETSc model required more than two years. However, based on the improvement of the GCCOM performance, our team feels that the adoption of the PETSc framework has been worth the effort.

Additionally, we find the PETSc based model to be portable: we recently successfully completed a prototype migration to the SDSC Comet system. The model ran to completion, but there is still much work to be done as we explore the optimal memory and core configuration. Another motivation to move GCCOM to a system like Comet is that they have access to optimized parallel IO libraries and file systems (such as the parallel NetCDF, and the Lustre system).

The current version of the parallel framework can be improved in several ways. Domain decomposition can be modified to take advantage of full partitioning in all three dimensions. However, this would require changing, or replacing, the existing pressure-gradient algorithm, which forces a vertical-slab decomposition because of self-recurrence in the spline integration over the column. The parallel model would also benefit from the use of parallel file input and output and improvements in memory management. We plan to explore how the adoption of exascale software systems such as ADIOS, which manages data between nodes in parallel with the computations on the nodes, will benefit the GCCOM model [53].

In conclusion, the performance tests conducted in these experiments show that the PETSc-based parallel GCCOM model satisfies several of its primary goals, including:

- Produce results that agree with the validated serial model,
- Decrease the time to solution while showing strong scalability,
- Deliver reproducible results that are not affected by data distribution across multiple nodes and cores,
- Maintain an efficiency that scales to several hundred cores without showing any signs of slowing down, and
- Establish a model that is portable and can operate in heterogeneous environments.

The results shown in this paper also show that GCCOM is a parallel and scaleable, multiphysics, and multiscale model: it scales to hundreds of cores (the limit of the test system); it can capture different types of phenomena, including fluid flow, nonhydrostatic pressure, thermodynamics, sea surface height; and it can operate over physical scales that range from 10^0 to 10^3 m.

Author Contributions: Conceptualization, M.V., M.G., and J.E.C.; methodology, M.V., M.G., M.P.T., and J.E.C.; software, M.V., M.G. and M.P.T.; validation, M.V., M.G.; formal analysis, M.V. and M.P.T.; investigation, M.V.; resources, J.E.C.; data curation, M.V.; writing—original draft preparation, M.V., M.G., M.P.T.; writing—review and editing, M.V., M.P.T. and M.G.; visualization, M.V.; supervision, J.E.C.; project administration, J.E.C.; funding acquisition, J.E.C.

Funding: This research was supported by the Computational Science Research Center (CSRC) at San Diego State University (SDSU), the SDSU Presidents Leadership Fund, the CSU Council on Ocean Affairs, Science and Technology (COAST) Grant Development Program, and the National Science Foundation (OCI 0721656, CC-NIE 1245312, MRI Grant 0922702). Portions of this work used the Extreme Science and Engineering Discovery Environment (XSEDE), which is supported by National Science Foundation Grant No. ACI-1548562.

Acknowledgments: The authors are grateful for the contribution of the work done by Neelam Patel on the DMDA PETSc framework. We also want to acknowledge helpful conversations with Ryan Walter on the internal wave, l beam and lock release experiments for validation, Paul Choboter for his contribution to the development of the current GCCOM model, and Jame Otto for his helpful insights into parallelization and hardware characterization. This research was supported by the Computational Science Research Center (CSRC) at San Diego State University (SDSU), the SDSU Presidents Leadership Fund, the CSU Council on Ocean Affairs, Science and Technology (COAST) Grant Development Program, and the National Science Foundation (OCI 0721656, CC-NIE 1245312, MRI Grant 0922702). Portions of this work used the Extreme Science and Engineering Discovery Environment (XSEDE), which is supported by National Science Foundation Grant No. ACI-1548562.

Conflicts of Interest: The authors declare no conflict of interest.

Appendix A

Table A1. Max Root-mean-squared (RMS) error per variable when comparing Serial GCCOM vs. Parallel GCCOM Model from 1 to 240 processors across 12 nodes.

Nodes	Processors	D [g/cm^3]	T [°C]	p [bar]	u [m/s]	v [m/s]	w [m/s]
				Max RMS error per Variable			
12	240	1.22×10^{-13}	7.16×10^{-10}	1.03×10^{-05}	7.46×10^{-08}	7.47×10^{-08}	1.33×10^{-07}
11	220	1.22×10^{-13}	7.16×10^{-10}	1.03×10^{-05}	7.46×10^{-08}	7.47×10^{-08}	1.33×10^{-07}
10	200	1.22×10^{-13}	7.15×10^{-10}	1.03×10^{-05}	7.45×10^{-08}	7.47×10^{-08}	1.33×10^{-07}
9	180	1.22×10^{-13}	7.16×10^{-10}	1.03×10^{-05}	7.46×10^{-08}	7.47×10^{-08}	1.33×10^{-07}
8	160	1.22×10^{-13}	7.16×10^{-10}	1.03×10^{-05}	7.46×10^{-08}	7.48×10^{-08}	1.33×10^{-07}
7	140	1.22×10^{-13}	7.16×10^{-10}	1.04×10^{-05}	7.46×10^{-08}	7.48×10^{-08}	1.33×10^{-07}
6	120	1.22×10^{-13}	7.16×10^{-10}	1.03×10^{-05}	7.46×10^{-08}	7.47×10^{-08}	1.33×10^{-07}
5	100	1.22×10^{-13}	7.16×10^{-10}	1.04×10^{-05}	7.46×10^{-08}	7.48×10^{-08}	1.33×10^{-07}
4	80	1.22×10^{-13}	7.15×10^{-10}	1.04×10^{-05}	7.55×10^{-08}	7.47×10^{-08}	1.33×10^{-07}
3	60	1.22×10^{-13}	7.15×10^{-10}	1.04×10^{-05}	7.42×10^{-08}	7.46×10^{-08}	1.33×10^{-07}
2	40	1.22×10^{-13}	7.14×10^{-10}	1.04×10^{-05}	7.41×10^{-08}	7.46×10^{-08}	1.33×10^{-07}
1	20	1.22×10^{-13}	7.15×10^{-10}	1.04×10^{-05}	7.42×10^{-08}	7.45×10^{-08}	1.33×10^{-07}
1	1	1.22×10^{-13}	7.13×10^{-10}	1.04×10^{-05}	7.35×10^{-08}	7.44×10^{-08}	1.33×10^{-07}

Table A2. Max RMS per variable when comparing Parallel GCCOM Model outputs from 20 to 240 processors across 12 nodes vs. Parallel GCCOM output in a single processor.

Nodes	Processors	D [g/cm^3]	T [°C]	p [bar]	u [m/s]	v [m/s]	w [m/s]
				Max RMS per Variable			
12	240	1.88×10^{-15}	9.96×10^{-12}	6.32×10^{-08}	2.73×10^{-09}	1.36×10^{-09}	7.91×10^{-10}
11	220	1.88×10^{-15}	9.98×10^{-12}	2.45×10^{-08}	2.76×10^{-09}	1.37×10^{-09}	8.16×10^{-10}
10	200	1.88×10^{-15}	1.00×10^{-11}	5.43×10^{-08}	2.78×10^{-09}	1.36×10^{-09}	8.01×10^{-10}
9	180	1.88×10^{-15}	1.03×10^{-11}	3.61×10^{-08}	2.49×10^{-09}	1.38×10^{-09}	8.53×10^{-10}
8	160	2.04×10^{-15}	1.04×10^{-11}	4.82×10^{-08}	2.64×10^{-09}	1.39×10^{-09}	8.69×10^{-10}
7	140	2.04×10^{-15}	1.07×10^{-11}	2.49×10^{-08}	2.46×10^{-09}	1.42×10^{-09}	9.22×10^{-10}
6	120	2.04×10^{-15}	1.04×10^{-11}	6.56×10^{-08}	2.64×10^{-09}	1.39×10^{-09}	8.57×10^{-10}
5	100	2.04×10^{-15}	1.07×10^{-11}	5.00×10^{-08}	2.72×10^{-09}	1.41×10^{-09}	9.06×10^{-10}
4	80	1.57×10^{-15}	8.07×10^{-12}	2.73×10^{-08}	2.12×10^{-09}	1.22×10^{-09}	6.93×10^{-10}
3	60	1.10×10^{-15}	5.46×10^{-12}	1.66×10^{-08}	1.54×10^{-09}	8.34×10^{-10}	4.57×10^{-10}
2	40	9.42×10^{-16}	4.13×10^{-12}	3.88×10^{-08}	1.05×10^{-09}	5.47×10^{-10}	3.47×10^{-10}
1	20	4.71×10^{-16}	1.84×10^{-12}	4.72×10^{-08}	1.54×10^{-09}	5.96×10^{-10}	1.68×10^{-10}

Table A3. Parallel performance of the high resolution stratified seamount experiment ($3000 \times 200 \times 100$) in the COD system.

Nodes	Processors	WTime [s]	I/O Main Loop [s]	WTime $-$ I/O [s]	Speedup	Efficiency	Streams Eff.
12	240	2.95×10^{02}	1.19×10^{02}	1.75×10^{02}	8.82×10^{01}	3.67×10^{-01}	3.17×10^{-01}
11	220	3.04×10^{02}	1.15×10^{02}	1.88×10^{02}	8.21×10^{01}	3.73×10^{-01}	3.16×10^{-01}
10	200	3.23×10^{02}	1.15×10^{02}	2.08×10^{02}	7.43×10^{01}	3.72×10^{-01}	3.13×10^{-01}
9	180	3.43×10^{02}	1.13×10^{02}	2.30×10^{02}	6.74×10^{01}	3.74×10^{-01}	3.15×10^{-01}
8	160	3.68×10^{02}	1.12×10^{02}	2.56×10^{02}	6.04×10^{01}	3.77×10^{-01}	3.16×10^{-01}
7	140	4.01×10^{02}	1.14×10^{02}	2.86×10^{02}	5.40×10^{01}	3.86×10^{-01}	3.14×10^{-01}
6	120	4.51×10^{02}	1.14×10^{02}	3.37×10^{02}	4.59×10^{01}	3.83×10^{-01}	3.16×10^{-01}
5	100	5.10×10^{02}	1.10×10^{02}	4.00×10^{02}	3.87×10^{01}	3.87×10^{-01}	3.12×10^{-01}
4	80	5.67×10^{02}	1.02×10^{02}	4.65×10^{02}	3.33×10^{01}	4.16×10^{-01}	3.11×10^{-01}
3	60	7.40×10^{02}	9.80×10^{01}	6.42×10^{02}	2.41×10^{01}	4.02×10^{-01}	3.14×10^{-01}
2	40	9.57×10^{02}	8.41×10^{01}	8.73×10^{02}	1.77×10^{01}	4.43×10^{-01}	3.10×10^{-01}
1	20	1.78×10^{03}	8.23×10^{01}	1.70×10^{03}	9.12×10^{00}	4.56×10^{-01}	3.07×10^{-01}
1	1	1.55×10^{04}	3.96×10^{01}	1.55×10^{04}	1.00	1.00	1.00

Table A4. Parallel performance of the medium-sized stratified seamount experiment ($2000 \times 100 \times 100$) in the COD system.

Nodes	Processors	WTime [s]	I/O Main Loop [s]	WTime $-$ I/O [s]	Speedup	Efficiency	Comm. Time	I/O Time [s]
12	240	1.05×10^{02}	5.60×10^{01}	4.91×10^{01}	7.02×10^{01}	2.93×10^{-01}	8.63×10^{06}	1.11×10^{02}
8	160	1.24×10^{02}	5.08×10^{01}	7.33×10^{01}	4.70×10^{01}	2.94×10^{-01}	4.03×10^{01}	1.07×10^{02}
4	80	1.77×10^{02}	4.27×10^{01}	1.34×10^{02}	2.57×10^{01}	3.21×10^{-01}	1.08×10^{01}	1.12×10^{02}
2	40	2.57×10^{02}	2.25×10^{01}	2.34×10^{02}	1.47×10^{01}	3.68×10^{-01}	3.59×10^{00}	9.68×10^{01}
1	1	3.46×10^{03}	1.22×10^{01}	3.45×10^{03}	1.00	1.00	1.00×10^{-05}	8.64×10^{02}

Table A5. Parallel performance of the low resolution stratified seamount experiment ($1500 \times 100 \times 50$) in the COD system.

Nodes	Processors	WTime [s]	I/O Main Loop [s]	WTime $-$ I/O [s]	Speedup	Efficiency	Comm. Time	I/O Time [s]
12	240	3.54×10^{01}	2.31×10^{01}	1.23×10^{01}	6.84×10^{01}	2.85×10^{-01}	6.84×10^{-04}	3.54×10^{01}
8	160	3.79×10^{01}	2.12×10^{01}	1.67×10^{01}	5.05×10^{01}	3.16×10^{-01}	5.05×10^{-04}	3.79×10^{01}
4	80	5.00×10^{01}	1.76×10^{01}	3.24×10^{01}	2.60×10^{01}	3.25×10^{-01}	2.60×10^{-04}	5.00×10^{01}
2	40	7.43×10^{01}	8.58×10^{00}	6.57×10^{01}	1.28×10^{01}	3.21×10^{-01}	1.28×10^{-04}	7.43×10^{01}
1	1	8.46×10^{02}	3.34×10^{00}	8.43×10^{02}	1.00	1.00	1.00×10^{-05}	8.46×10^{02}

References

1. Alowayyed, S.; Groen, D.; Coveney, P.V.; Hoekstra, A.G. Multiscale computing in the exascale era. *J. Comput. Sci.* **2017**, *22*, 15–25. [CrossRef]
2. Abouali, M.; Castillo, J.E. Unified Curvilinear Ocean Atmosphere Model (UCOAM): A vertical velocity case study. *Math. Comput. Model.* **2013**, *57*, 2158–2168. [CrossRef]
3. Marshall, J.; Jones, H.; Hill, C. Efficient ocean modeling using non-hydrostatic algorithms. *J. Mar. Syst.* **1998**, *18*, 115–134. [CrossRef]
4. Fringer, O.B.; McWillimas, J.C.; Street, R.L. A New Hybrid Model for Coastal Simulations. *Oceanography* **2006**, *19*, 64–77. [CrossRef]
5. Berntsen, J.; Xing, J.; Davies, A.M. Numerical studies of flow over a sill: Sensitivity of the non-hydrostatic effects to the grid size. *Ocean Dyn.* **2009**, *59*, 1043–1059. [CrossRef]
6. Torres, C.R.; Hanazaki, H.; Ochoa, J.; Castillo, J.; Van Woert, M. Flow past a sphere moving vertically in a stratified diffusive fluid. *J. Fluid Mech.* **2000**, *417*, 211–236. [CrossRef]
7. Marshall, J.; Adcroft, A.; Hill, C.; Perelman, L.; Heisey, C. A finite-volume, incompressible Navier–Stokes model for studies of the ocean on parallel computers. *J. Geophys. Res. Oceans* **1997**, *102*, 5753–5766. [CrossRef]
8. Fringer, O.B.; Gerritsen, M.; Street, R.L. An unstructured-grid, finite-volume, nonhydrostatic, parallel coastal ocean simulator. *Ocean Model.* **2006**, *14*, 139–173. [CrossRef]
9. Lai, Z.; Chen, C.; Cowles, G.W.; Beardsley, R.C. A nonhydrostatic version of FVCOM: 1. Validation experiments. *J. Geophys. Res. Oceans* **2010**, *115*, 1–23. [CrossRef]

10. Santilli, E.; Scotti, A. The stratified ocean model with adaptive refinement (SOMAR). *J. Comput. Phys.* **2015**, *291*, 60–81. [CrossRef]

11. Liu, Z.; Lin, L.; Xie, L.; Gao, H. Partially implicit finite difference scheme for calculating dynamic pressure in a terrain-following coordinate non-hydrostatic ocean model. *Ocean Model.* **2016**, *106*, 44–57. [CrossRef]

12. Garcia, M.; Choboter, P.F.; Walter, R.K.; Castillo, J.E. Validation of the nonhydrostatic General Curvilinear Coastal Ocean Model (GCCOM) for stratified flows. *J. Comput. Sci.* **2019**, *30*, 143–156. [CrossRef]

13. Thomas, M.P.; Castillo, J.E. Parallelization of the 3D Unified Curvilinear Coastal Ocean Model: Initial Results. In Proceedings of the International Conference on Computational Science and Its Applications, Amsterdam, The Netherlands, 4–6 June 2012; pp. 88–96.

14. Thomas, M.P.; Bucciarelli, R.; Chao, Y.; Choboter, P.; Garcia, M.; Manjunanth, S.; Castillo, J.E. Development of an Ocean Sciences Education Portal for Simulating Coastal Ocean Processes. In Proceedings of the 10th Gateway Computing Environments Workshop, Boulder, CO, USA, 29–31 September 2015.

15. Choboter, P.F.; Garcia, M.; Cecchis, D.D.; Thomas, M.; Walter, R.K.; Castillo, J.E. Nesting nonhydrostatic GCCOM within hydrostatic ROMS for multiscale Coastal Ocean Modeling. In Proceedings of the MTS IEEE Oceans 2016 Conference, Monterey, CA, USA, 19–23 September 2016; pp. 1–4. [CrossRef]

16. Balay, S.; Abhyankar, S.; Adams, M.F.; Brown, J.; Brune, P.; Buschelman, K.; Dalcin, L.; Eijkhout, V.; Gropp, W.D.; Kaushik, D.; et al. *PETSc Users Manual*; Technical Report ANL-95/11—Revision 3.8; Argonne National Laboratory: Lemont, IL, USA, 2017.

17. Balay, S.; Abhyankar, S.; Adams, M.F.; Brown, J.; Brune, P.; Buschelman, K.; Dalcin, L.; Eijkhout, V.; Gropp, W.D.; Kaushik, D.; et al. PETSc Web Page, 2018. Available online: http://www.mcs.anl.gov/petsc (accessed on 12 June 2019).

18. Balay, S.; Gropp, W.D.; McInnes, L.C.; Smith, B.F. Efficient Management of Parallelism in Object Oriented Numerical Software Libraries. In *Modern Software Tools in Scientific Computing*; Arge, E., Bruaset, A.M., Langtangen, H.P., Eds.; Birkhäuser Press: Basel, Switzerland, 1997; pp. 163–202.

19. Valera, M.; Garcia, M.; Walter, M.; Choboter, R.; Castillo, P. Modeling nearshore internal bores and waves in Monterey bay using the General Curvilinear Coastal Ocean Dynamics Model, GCCOM. In Proceedings of the Joint AMS-SIAM Meeting, Minisymposium on Mimetic Multiphase Subsurface and Oceanic Transport, San Diego, CA, USA, 10–13 January 2018.

20. Garcia, M.; Hoar, T.; Thomas, M.; Bailey, B.; Castillo, J. Interfacing an ensemble Data Assimilation system with a 3D nonhydrostatic Coastal Ocean Model, an OSSE experiment. In Proceedings of the MTS IEEE Oceans 2016 Conference, Monterey, CA, USA, 19–23 September 2016; pp. 1–11. [CrossRef]

21. Shchepetkina, A.; McWilliamsa, J.C. Algorithm for non-hydrostatic dynamics in the Regional Oceanic Modeling System. *Ocean Model.* **2007**, *18*, 143–174.

22. Kerbyson, D.J.; Jones, P.W. A Performance Model of the Parallel Ocean Program. *Int. J. High Perform. Comput. Appl.* **2005**, *19*, 261–276. [CrossRef]

23. Ringler, T.; Petersen, M.; Higdon, R.L.; Jacobsen, D.; Jones, P.W.; Maltrud, M. A multi-resolution approach to global ocean modeling. *Ocean Model.* **2013**, *69*, 211–232. [CrossRef]

24. Chassignet, E.P.; Hurlburt, H.E.; Smedstad, O.M.; Halliwell, G.R.; Hogan, P.J.; Wallcraft, A.J.; Baraille, R.; Bleck, R. The HYCOM (HYbrid Coordinate Ocean Model) data assimilative system. *J. Mar. Syst.* **2007**, *65*, 60–83. [CrossRef]

25. Tang, H.; Qu, K.; Wu, X. An overset grid method for integration of fully 3D fluid dynamics and geophysics fluid dynamics models to simulate multiphysics coastal ocean flows. *J. Comput. Phys.* **2014**, *273*, 548–571. [CrossRef]

26. Zijlema, M.; Stelling, G.; Smit, P. SWASH: An operational public domain code for simulating wave fields and rapidly varied flows in coastal waters. *Coastal Eng.* **2011**, *58*, 992–1012. [CrossRef]

27. Brzenski, J. Coupling GCCOM, a Curvilinear Ocean Model Rigid Lid Simulation with SWASH for Analysis of Free Surface Conditions. Master's Thesis, San Diego State University, San Diego, CA, USA, 2019.

28. Torres, C.R.; Castillo, J.E. A New 3D Curvilinear Coordinates Numerical Model for Oceanic Flow Over Arbitrary Bathymetry (In Spanish). In Proceedings of the Desarrollos Recientes en Métodos Numéricos, Muller-Karger, C.M., Lentini, M., Cerrolaza, M., Eds.; 2002; pp. 105–112, ISBN 980-00-1951-0. Available online: https://www.researchgate.net/publication/233853889_A_new_3d_curvilinear_coordinates_numerical_model_for_oceanic_flow_over_arbitrary_bathymetry_In_Spanish (accessed on 13 June 2019).

29. Torres, C.R.; Castillo, J.E. Stratified Rotating Flow Over Complex Terrain. *Appl. Numer. Math.* **2003**, *47*, 531–541. [CrossRef]

30. Torres, C.R.; Mascarenhas, A.S.; Castillo, J.E. Three-dimensional stratified flow over the Alarcón Seamount, Gulf of California entrance. *Deep Sea Res. Part II Top. Stud. Oceanogr.* **2004**, *51*, 647–657. [CrossRef]

31. Abouali, M.; Castillo, J.E. *General Curvilinear Ocean Model (GCOM) Next, Generation*; Technical Report CSRCR2010-02; Computational Sciences Research Center, San Diego State University: San Diego, CA, USA, 2010.

32. Arakawa, A. Computational Design for Long–Term Numerical Integration of the Equations of Fluid Motion: Two–Dimensional Incompressible Flow. Part I. *J. Comput. Phys.* **1997**, *135*, 103–114. [CrossRef]

33. Garcia, M. Data Assimilation Unit for the General Curvilinear Environmental Model. Ph.D. Dissertation, Claremont Graduate University, Claremont, CA, USA; San Diego State University, San Diego, CA, USA, 2015.

34. Notay, Y. *User's Guide to AGMG*, 2014; Technical Report; Service de Metrologie Nucleaire, Universite Libre de Bruxelles: Brussels, Belgium, 2014. Available online: http://agmg.eu/agmg_userguide.pdf (accessed on 13 June 2019).

35. Thomas, M.P. Parallel Implementation of the Unified Curvilinear Ocean and Atmospheric (UCOAM) Model and Supporting Computational Environment. Ph.D. Dissertation, Claremont Graduate University, Claremont, CA, USA; San Diego State University, San Diego, CA, USA, 2014.

36. Smith, B.; McInnes, L.C.; Constantinescu, E.; Adams, M.; Balay, S.; Brown, J.; Knepley, M.; Zhang, H. PETSc's Software Strategy for the Design Space of Composable Extreme–Scale Solvers. In Proceedings of the OE Exascale Research Conference, Portland, OR, USA, 16–19 July 2012; pp. 1–8.

37. Valera, M.; Patel, N.; Castillo, J. *PETSc-Based Parallelization of the fully 3D-Curvilinear Non–Hydrostatic Coastal Ocean Dynamics Model, GCCOM*; Technical Report CSRCR2017-02; Computational Sciences Research Center, San Diego State University: San Diego, CA, USA, 2017.

38. Patel, N.V. Validation of a PETSc-Based Parallel General Curvilinear Coastal Ocean Model. Master's Thesis, San Diego State University, San Diego, CA, USA, 2017.

39. Wang, W.; Fischer, T.; Zehner, B.; Böttcher, N.; Görke, U.J.; Kolditz, O. A parallel finite element method for two-phase flow processes in porous media: OpenGeoSys with PETSc. *Environ. Earth Sci.* **2015**, *73*, 2269–2285. [CrossRef]

40. Liu, L.; Li, R.; Yang, G.; Wang, B.; Li, L.; Pu, Y. Improving parallel performance of a finite-difference AGCM on modern high-performance computers. *J. Atmos. Ocean. Technol.* **2014**, *31*, 2157–2168. [CrossRef]

41. Brown, J.; Knepley, M.G.; May, D.A.; McInnes, L.C.; Smith, B. Composable linear solvers for multiphysics. In Proceedings of the 2012 11th International Symposium Parallel Distributed Computing ISPDC, Munich, Germany, 25–29 June 2012; pp. 55–62. [CrossRef]

42. Shi, J.; Li, R.; Xi, Y.; Saad, Y.; de Hoop, M.V. Computing Planetary Interior Normal Modes with a Highly Parallel Polynomial Filtering Eigensolver. In Proceedings of the SC18: International Conference for High Performance Computing, Networking, Storage and Analysis, Dallas, TX, USA, 11–16 November 2018; pp. 894–906. [CrossRef]

43. May, D.A.; Sanan, P.; Rupp, K.; Knepley, M.G.; Smith, B.F. Extreme-Scale Multigrid Components within PETSc. In Proceedings of the Platform for Advanced Scientific Computing Conference (PASC '16), New York, NY, USA, 8–10 June 2016; pp. 1–24. [CrossRef]

44. Heroux, M.A.; Bartlett, R.A.; Howle, V.E.; Hoekstra, R.J.; Hu, J.J.; Kolda, T.G.; Lehoucq, R.B.; Long, K.R.; Pawlowski, R.P.; Phipps, E.T.; et al. An overview of the Trilinos project. *ACM Trans. Math. Softw.* **2005**, *31*, 397–423. [CrossRef]

45. Falgout, R.D.; Jones, J.E.; Yang, U.M. The Design and Implementation of hypre, a Library of Parallel High Performance Preconditioners. In *Numerical Solution of Partial Differential Equations on Parallel Computers*; Bruaset, A., Tveito, A., Eds.; Springer: Berlin, Germany, 2006; pp. 267–294.

46. Lamb, C. OpenCL for NVIDIA GPUs. In Proceedings of the 2009 IEEE Hot Chips 21 Symposium (HCS), Stanford, CA, USA, 23–25 August 2009; pp. 1–24. [CrossRef]

47. Shchepetkin, A.F.; McWilliams, J.C. The regional oceanic modeling system (ROMS): A split-explicit, free-surface, topography-following-coordinate oceanic model. *Ocean Model.* **2005**, *9*, 347–404. [CrossRef]

48. Boussinesq, J. *ThÉorie de l'Écoulement Tourbillonnant et Tumultueux des Liquides dans les Lits Rectilignes a Grande Section*; Gauthier-Villars et fils: Paris, France, 1897; p. 94.

49. Moore, R.L.; Tatineni, M.; Wagner, R.P.; Wilkins-Diehr, N.; Norman, M.L.; Baru, C.; Baxter, D.; Fox, G.C.; Majumdar, A.; Papadopoulos, P.; et al. Gateways to Discovery: Cyberinfrastructure for the Long Tail of Science. In Proceedings of the 2014 Annual Conference on Extreme Science and Engineering Discovery Environment, Atlanta, GA, USA, 13–18 July 2014. [CrossRef]
50. Zender, C.S. Bit Grooming: Statistically accurate precision-preserving quantization with compression, evaluated in the netCDF Operators (NCO, v4.4.8+). *Geosci. Model Dev.* **2016**, *9*, 3199–3211. [CrossRef]
51. Pacheco, P. *An Introduction to Parallel Programming*; Morgan Kaufmann: Burlington, MA, USA, 2011; p. 392.
52. Foster, I. *Designing and Building Parallel Programs: Concepts and Tools for Parallel Software Engineering*; Addison–Wesley Publishing Company: Boston, MA, USA, 1995.
53. Liu, Q.; Logan, J.; Tian, Y.; Abbasi, H.; Podhorszki, N.; Choi, J.Y.; Klasky, S.; Tchoua, R.; Lofstead, J.; Oldfield, R.; et al. Hello ADIOS: The Challenges and Lessons of Developing Leadership Class I/O Frameworks. *Concurr. Comput. Pract. Exp.* **2014**, *26*, 1453–1473. [CrossRef]

Journal of
*Marine Science
and Engineering*

Review

Modeling Multiscale and Multiphysics Coastal Ocean Processes: A Discussion on Necessity, Status, and Advances

Hansong Tang [1,*], Charles Reid Nichols [2], Lynn Donelson Wright [3] and Donald Resio [4]

1   Civil Engineering Department, City College of New York, CUNY, New York, NY 10031, USA
2   Marine Information Resources Corporation, 340 Wye Narrows Drive, Queenstown, MD 21658, USA; rnichols@mirc-us.com
3   Southeastern Universities Research Association, Washington, DC 20005, USA; wright@sura.org
4   Taylor Engineering Research Institute, University of North Florida, Jacksonville, FL 32224, USA; don.resio@unf.edu
*   Correspondence: htang@ccny.cuny.edu

Abstract: Coastal ocean flows are interconnected by a complex suite of processes. Examples are inlet jets, river mouth effluents, ocean currents, surface gravity waves, internal waves, wave overtopping, and wave slamming on coastal structures. It has become necessary to simulate such oceanographic phenomena directly and simultaneously in many disciplines, including coastal engineering, environmental science, and marine science. Oceanographic processes exhibit distinct behaviors at specific temporal and spatial scales, and they are multiscale, multiphysics in nature; these processes are described by different sets of governing equations and are often modeled individually. In order to draw the attention of the scientific community and promote their simulations, a Special Issue of the *Journal of Marine Science and Engineering* entitled "Multiscale, Multiphysics Modelling of Coastal Ocean Processes: Paradigms and Approaches" was published. The papers collected in this issue cover physical phenomena, such as wind-driven flows, coastal flooding, turbidity currents, and modeling techniques such as model comparison, model coupling, parallel computation, and domain decomposition. This article outlines the needs for modeling of coastal ocean flows involving multiple physical processes at different scales, and it discusses the implications of the collected papers. Additionally, it reviews the current status and offers a roadmap with numerical methods, data collection, and artificial intelligence as future endeavors.

Keywords: multiscale; multiphysics; model coupling; domain decomposition; data collection; machine learning

Citation: Tang, H.; Nichols, C.R.; Wright, L.D.; Resio, D. Modeling Multiscale and Multiphysics Coastal Ocean Processes: A Discussion on Necessity, Status, and Advances. *J. Mar. Sci. Eng.* **2021**, *9*, 847. https://doi.org/10.3390/jmse9080847

Academic Editor: Alvise Benetazzo

Received: 9 June 2021
Accepted: 28 July 2021
Published: 5 August 2021

Publisher's Note: MDPI stays neutral with regard to jurisdictional claims in published maps and institutional affiliations.

1. Background and Necessities

As a consequence of environmental change and ever-expanding human activities, it has become urgently needed to investigate many emerging oceanic flow problems. Three examples manifest such needs. An anthropogenic example was the 2010 Gulf of Mexico oil spill, which started as a jet at the seafloor and rose to become floating oil patches that led to an environmental disaster [1]. Overbank compound flooding by two North Carolina coastal rivers provided an example of a complex suite of natural hazards that resulted from ocean surges, fluvial waters, and inland runoff, plus their interactions, during hurricanes Dennis and Floyd [2]. An engineering example is illustrated by the fast-growing, worldwide practice of power generation from ocean current energy, in which both local flows at turbines and the background tides play a role [3]. The study of these problems has significant impacts on advances in sciences, engineering, and resilient coastal communities. Towards the study, various programs have been established, such as Southeastern Universities Research Association (SURA), Coastal Ocean Observing and Prediction Program (SCOOP), and the NOAA Coastal and Ocean Modeling Testbed program [4].

The three example problems listed above have differing characteristics, but all of them bear a similar feature: they are multiscale and multiphysics in nature and present challenges to today's modeling capabilities. For instance, the BP oil spill in the Gulf of Mexico started as a fully three-dimensional (3D), high-speed jet with intense mixing, at scales on the order of 10 m, and later it evolved into drifting patches of oil film, with scales on the order of 100 km horizontally. Currently, many models have been developed in the ocean science community for applications such as circulation, surges, and waves [5–8]. However, these models lack appropriate capabilities to directly account for multiscale, multiphysics phenomena, particularly those fully 3D, local, complex phenomena, such as the dynamic processes in the initial jet in the BP oil spill and water splashing as shown in Figure 1. Without a multiscale and multiphysics approach, these models could only partially simulate critical processes in those emerging problems. For instance, there is currently no single model or software package that can directly simulate the whole process from the jet all the way to the floating oil in the BP oil spill case.

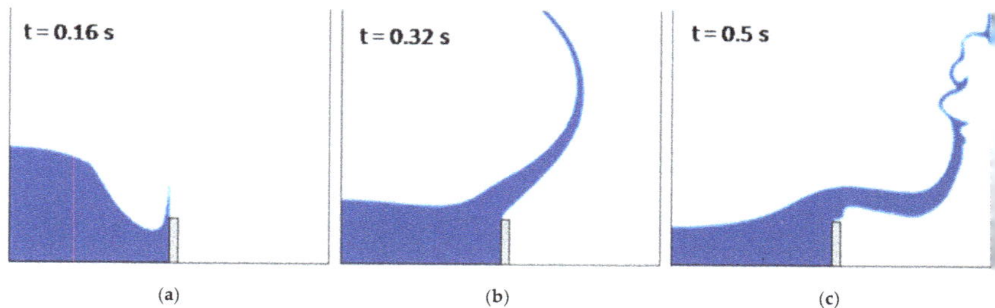

Figure 1. Simulated collapse of a water column, on the left, and its slamming on a plate, in the middle, and its splashing at a wall, on the right [9]. The simulation is produced by a solver for the Navier–Stokes equations.

For many years, oceanographers have worked towards simulating multiscale coastal ocean flows, and now it has become a common practice within the ocean science community [10–12]. For these simulations, meshes are refined locally, either via stretched meshes or nested meshes at local regions, so that not only the background large-scale flow patterns are captured, but also fine-scale, local motions are resolved. However, this is not the case for multiphysics simulations, which in general cannot be realized merely by local mesh refinement. Traditionally, multiphysics refers to a system with multiple phenomena, for instance, thermal diffusion, fluid flow, and phase change [13,14]. Even though these phenomena are interdependent, they are described by different governing equations. In the ocean, it is typical for a flow problem to involve multiple processes that exhibit distinct physical behaviors, e.g., a jet at the seafloor and floating oil at the water surface in the BP *Deepwater Horizon* Oil Spill. These processes are better described by different governing equations, and frequently they are multiscale in nature owing to a vast range of spatial and temporal scales. Therefore, they are referred to as multiphysics flows [15–17], although such terminology is not generally recognized in the ocean science community.

2. A Discussion of the Collected Papers

In order to promote the simulations of multiscale and multiphysics coastal ocean flows, a Special Issue of the *Journal of Marine Science and Engineering* entitled "Multiscale, Multiphysics Modelling of Coastal Ocean Processes: Paradigms and Approaches" was initiated [18]. This Special Issue collected several papers, each of which focused on a specific topic. Their topics included flooding, effects of scales and wind fields, model assessment, model coupling, parallel computation, and computational methods. Although small in numbers, the collected papers exemplified recent main efforts in the simulations of multiscale, multiphysics flow problems. The research papers are reviewed in the paragraphs that follow.

It is common to conduct multiscale simulations by increasing mesh resolution and downscaling to resolve small-scale flow events such as flooding in the streets. When such a simulation goes into local regions, it is crucial to assure its accuracy and reliability because hydrodynamic phenomena and their interaction with the environmental settings, such as buildings, become very complicated. An indispensable step is to validate the models via abundant data that characterize the local region. Spatial complexity near the coast could be much greater than in the deep ocean where data coverage from observations and satellites requires less spatial detail. However, it is challenging to collect sufficient data during extreme coastal events, such as street-level data for swift-flowing water during storms. An innovative effort to collect street flooding data (e.g., high water marks) during two consecutive storms involved observations over a thousand local residents at Hampton Roads, VA in an activity called 'Catch the King' [19]. 'Catch the King' was well received among residents, who were educated and trained on data collection. After being processed, the data were used to calibrate the model for the VIMS' Tidewatch storm tide inundation maps. Such work is not only novel and effective to better capture local flooding but also increases the awareness of residents, bearing on a broader social impact.

In a simulation to resolve local flows, adopting appropriate models has been another critical issue besides mesh refinement along coastlines. As a result, a comparison of model performance has become necessary [20]. Driven by the need for improved marine safety and emergency response, the performance of NEMO and FVCOM was tested through scientific collaboration with multiple teams [21]. NEMO and FVCOM are two distinct types of models; the former uses a finite difference method on a structured mesh, while the latter adopts a finite volume method on an unstructured mesh. The study area is the Saint John Harbor in the Bay of Fundy, which features a complex flow system of waters from the ocean and rivers. The system exhibits intricate patterns at different scales, and its simulation is a nontrivial test for both models. The mesh resolution at the coast is as fine as about 100 m, but the authors anticipated that both models may reach their limits if it gets further fined. The authors concluded that overall, the two models performed similarly in accuracy compared to field observation, and FVCOM has a smaller computational cost. Generally speaking, an unstructured-grid model requires more computational time, but FVCOM can deal with irregular coastlines with a smaller number of grid nodes and, which may lead to a lower computational cost. This feature has been exploited to support wave energy survey along vast portions of the coastal ocean [22].

The multiscale, multiphysics nature of coastal ocean flows is attributed to various factors, and wind fields are among those that play an essential role. With the aid of FVCOM, a comparison study was made on effects from cold fronts in 2014 and Hurricane Barry in 2019 on flow patterns inside Barataria Bay [23]. In general, these fronts and the hurricane's wind fields exhibited distinct differences not only in temporal and spatial scales but also in directions. Such differences lead to an interesting disparity in behaviors of the hydrodynamics in the bay. For instance, after the passage of a cold wind front, the water surface inside the bay presents a trough, while it exhibits surges after the hurricane. As a result, the study showed that water was transported out of the bay after a cold front of winds passes, whereas it is pushed into the bay after Hurricane Barry's landfall. This study also indicated that FVCOM was able to capture flow patterns inside a bay driven by wind fields at different scales.

Various complex flow behaviors result from the multiscale, multiphysics nature of coastal ocean flows. Extreme atmospheric wind and precipitation have contributed to unusual flooding along two rivers located along the NC coast during Hurricanes Dennis and Floyd [2]. The compound flooding events resulted from storm surges and heavy rainfall. They are each complicated phenomena but exacerbated by the simultaneous impact of the two hurricanes, resulting in interactions that contributed to dangerous flooding events. Based on data from observation and their 1D modeling, the authors delved into surface elevation and flow rates in the rivers, surge heights in the ocean, etc. They presented a clear description of the mechanism of the downstream blocking during the flooding. Importantly, this paper indicated that, due to strong interaction among phenomena at different scales, existing modeling approaches are not appropriate because they are typically based on univariate methods and coarse resolution. This work revealed limitations in modeling and the need for multiscale and multiphysics modeling in coastal flooding, and it concluded that approaches based on coupling of multiple models should be adopted.

Simulation of flooding events through model coupling is becoming a trend, and research related to flows along the Gulf of Mexico's continental margin represents an effort across temporal scales, where the researchers coupled models and form a holistic modeling framework to capture turbidity currents at the seabed during storms [24]. The simulation involved various physical processes, such as fluvial flows, estuary currents, surface waves, and sediment transport, which behave differently and happen at different scales. The framework assembled component models for point, 2D, and 3D processes, primarily in the one-way coupling. It simulated these processes with resolution as fine as less than 1 s in time and 3 m in space. Based on a series of simulations and analyses of their results and actual data for hurricanes Gustav and Ike, a turbidity current problem as a representation of those in the Gulf of Mexico was formulated and simulated. The simulation revealed that hurricanes could bring a substantial amount of sand from coastal to deep waters. The authors' work dealt with complex processes, and it was more complicated than earlier studies on model problems, such as the motion of sand dunes due to surface waves [25].

Due to the inherent limitations in conventional coastal ocean models' governing equations, such as the hydrostatic assumption and parameterization, they cannot handle many complex local events, especially fully small-scale, 3D phenomena. Adoption of the Navier–Stokes equations is a remedy to overcome this problem [26], since, in principle, such equations can resolve all phenomena at different scales that are of interest. However, solving these equations is very expensive, and efficient computation is a huddle for moving forward. With such motivation, a full Navier–Stokes solver on a structured mesh was developed [27]. In this solver, the Fortran-interfaced Portable–Extensible Toolkit for Scientific Computation (PETSc) library equipped with domain decomposition techniques was utilized for parallel computation of the solver, particularly its Poisson equation for pressure. Because of the adoption of such parallelization, the increased speed of computation is substantial. The work demonstrates that enough mesh resolution is desired in capturing complex flow structures at a seamount, while the resolution can be reduced in the region far away from it. This paper provided a valuable addition to the sparse number of works on domain decomposition techniques for ocean flows.

In a broader sense, domain decomposition is an indispensable approach to achieving multiscale and multiphysics simulations. A theoretical study on domain decomposition and data assimilation in the computation of a linearized version of the shallow water equations was demonstrated for Baltic Sea circulation [28,29]. In this study, assimilation data with randomness (to mimic observation data) was imposed at an open boundary of a subdomain that was linked to another subdomain. Its computation was formulated into an inverse problem, whose objective function was built on the governing equations and boundary conditions with a term of the Tikhonov regularization [28]. Additionally, a discussion on uniqueness and computational steps was presented. In a numerical experiment on a model problem, the search in the optimization converges in a few steps to the assimilation data at the open boundary. Note that simplification, such as linearization or omitting the advection terms, is made, and the scenario of this study differs from realistic situations. Nevertheless, this work is particularly valuable since its topic and methods are novel, and publications on domain decomposition for ocean flows are infrequent.

The above-collected papers provide a perspective of typical current efforts to simulate multiscale, multiphysics flow problems. However, they only reflect a portion of the past efforts. For a more complete view on the current status of such simulation, a brief but more comprehensive review is presented about additional work on the theme in the following section.

3. Current Status

Many coastal ocean models have been built on geophysical fluid dynamics (GFD) equations in the past few decades. These models have successfully simulated various applications relevant to acoustics, ocean currents, surface waves, thermoclines, etc., and examples of them are POM, ADCIRC, ROMS, WAVEWATCH [5–8]. However, they cannot handle many emerging problems involving small-scale, complicated flows, such as the examples listed at the beginning of this article, which have largely been considered as secondarily important in the ocean science communities in the past decades. At the same time, many models have been developed in the engineering communities based on different equations, e.g., the Navier–Stokes equations. In principle and practice, they can directly simulate these small-scale, complex, local ocean flows of our interest, including those in the three examples mentioned above. Figure 1 shows samples of modeling of these small-scale, complex flows. Here, "directly" means without or with minimal simplifications and parametrizations that are commonly adopted in coastal ocean models,(e.g., drag coefficients of winds over water surfaces or at the seabed). Since the computation of the equations for the local flows, e.g., the Navier–Stokes equations, is very expensive, applying these engineering models to a large area of oceans could become prohibitive. In addition, such an approach may not be as efficient as coastal ocean models, for instance, for surface waves.

For more than a decade, simulations of multiscale ocean flows have been popular among the ocean science community. Within the frame of a conventional coastal ocean model, a general approach situation is to adopt multiple grid resolutions at different zones in the domain of computation. In attempts to simulate global ocean currents, simulations obtained with different mesh resolution indicate that finer resolution is indeed helpful to better resolve observed flow patterns, such as water surface elevation and flows through straights [30,31]. Fine resolution is frequently applied to nearshore regions to resolve various events there. For instance, to search best sites for marine kinetic hydrodynamic energy near coastlines, grid spacing less than 10 m is applied along the entire coast, while that over 10 m is used in open waters [32]. Three sets of nested grids with 3-arcminute resolution as the fine resolution are adopted in a wave energy survey in Indonesia waters [33]. Another event with high resolution in nearshore regions is coastal flooding, and grid spacing as fine as 3 m is adopted to resolve floods in streets [34]. A comparison is between a structured-grid model with nested-grids and an unstructured-grid with local mesh refinement, and it is concluded that the latter is more expensive in terms of computation [35]. It has become a common practice in the ocean science community to capture multiscale flow phenomena via local mesh refinement. Because of the complexity of the flows and their multiscale nature, it is not always straightforward to design multi-resolution meshes. For instance, a dense mesh at a tidal inlet could add dissipation to the solution there [36], and thus discretion is needed to achieve the desired accuracy there.

Multiphysics flows present more challenges, and the development of new modeling capabilities becomes necessary. Such flows tend to be associated with multiple temporal and spatial scales. However, they cannot be resolved simply by multi-resolution, or local mesh refinement, in the aforementioned conventional coastal ocean models. A direct approach is to build models on the basis of the Navier–Stokes equations or their variants in the whole computational domains [26,27]. In principle, such models, e.g., Fluidity-ICOM [26], are able to handle multiphysics flows beyond the reach of the conventional coastal ocean models, such as water slamming in Figure 1. However, like the aforementioned Navier–Stokes solvers developed in engineering communities, these models face difficulties, such as high computational cost, in application of ocean flows. Given the fact that models for individual phenomena at specific scales, e.g., large-scale ocean circulations and small-scale wave breaking, have become mature, and, as the most feasible and promising approach, it is natural to combine these models into a single framework to capture multiple physical phenomena, and such efforts start over a decade ago [15,37,38].

During the past decade, substantial research has focused on coupling the equations for surface waves and GFD equations (e.g., shallow water equations) for ocean currents, two typical oceanic phenomena. Examples cover the interaction between surface waves and ocean currents [39,40], wave-current-sediment motion [41], wave-driven morphology [25] effects of ocean currents on waves [42], and ice-induced wave attenuation [43]. It should be noted that waves and currents differ in temporal and spatial scales, and time steps for their computation are also distinct due to stability requirements [25,41]. Another type of coupling is between models for coastal ocean flows. Examples are coupling of a model for narrow tributaries with vertically 2D flow patterns to a model for 3D flows [44], a 2D Godunov-type model simulating local flooding across traffic roads to FVCOM for the background ocean currents [45], a shallow water flow solver and the Navier–Stokes solver to resolve local flows [46], and a 3D ocean model with fine grids on the order of a meter to ROMS with coarse grids [47]. As a most recent effort, solvers for the Navier–Stokes equations are coupled with FVCOM to simulate local, complex flows in high fidelity [48,49]. Figure 2 presents an example of this kind of coupling. In this case, the Navier–Stokes solver and FVCOM are state-of-the-art models used respectively by the engineering community and the ocean science community. Additional types of coupling include ocean circulation and sea ice [50], atmosphere, ocean, and biomaterials [51], storm surges and turbidity currents [24,46], and ocean surge and land runoff [52]. Other relevant efforts include the Earth System Modeling Framework (ESMF), which integrates many distinct geophysical models [53].

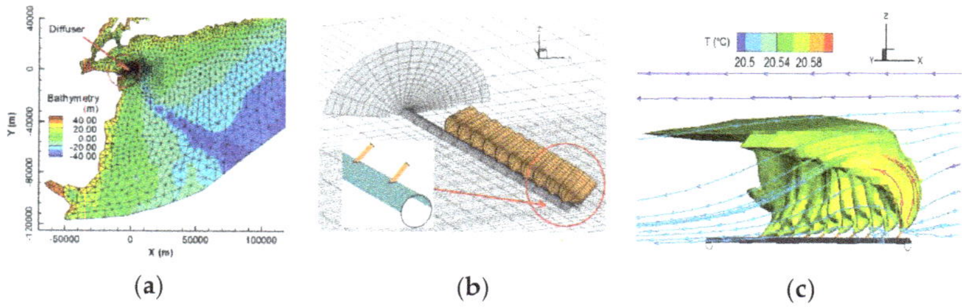

(a)   (b)   (c)

Figure 2. Simulation of thermal effluents discharged from a diffuser on seabed by coupling of a Navier–Stokes solver and FVCOM [16,48]. In the simulation, the former resolves local flows, and the latter captures the background currents, and the two models are coupled in two-way and march in time simultaneously. (**a**) Mesh of FVCOM. (**b**) Mesh of the Navier–Stokes equations solver. (**c**) Simulated thermal effluent at flood tide.

In general, coupling models to simulate multiphysics flows is a challenging complex task. Currently, the coupling is usually one-way, which is implemented by programs or manually, e.g., [24]. The one-way coupling may not only miss the feedback between solutions of different models but may also introduce substantial errors and uncertainties when passing solution data between models [54,55]. For instance, as in [55], a European group concluded that it is required to resolve 3D flow structures in near fields and the interaction between near and far fields to reduce such inaccuracy and uncertainty in tidal power development. As a result, efforts have been made to achieve two-way coupling, e.g., [45,48,49,56]. However, the bidirectional coupling is not yet widespread because it is challenging to realize and expensive to compute.

Efforts have been made on algorithm development and computational analysis to promote multiscale and multiphysics simulations. As another approach, the hybrid methods have been investigated, which merge different algorithms within a same model, rather than coupling different models as discussed above. For instance, equations for wave, current, and morphology are discretized and computed in a single system [25]. It is proposed to adopt a hybrid of continuous and discontinuous Galerkin methods to compute a generalized wave-continuity equation in ADCIRC [57]. In capturing internal waves, a method is proposed to combine a hydrostatic simulation on a grid and a non-hydrostatic simulation on another grid, plus a technique to switch between the two solutions [58]. A method that allows different time steps in different zones of the computational domain is presented to solve the shallow water equations [59]. Within ROMS, techniques are presented to realize two-way nesting to resolve local flows [60].In the past year, attempts have been made to analyze methods and computation, although such analyses are difficult and progress is limited due to the complexity of involved governing equations and also ocean flows. Research examples include optimal interface conditions to couple hydrostatic and nonhydrostatic ocean models [61], variational data assimilation [28], computational algorithms [62], and interface algorithms and stability analysis [63]. In a broad sense, all of these efforts fall into heterogeneous domain decomposition methods for coupling and computation with different partial differential equations, distinct numerical methods, and even dissimilar meshes [12].

4. Future Efforts

To move forward, important aspects that deserve efforts are mathematical foundation, algorithms, and computational power. In coupling different models, a crucial issue is to develop algorithms for computations at interfaces. Currently, interface treatments are ad hoc in theoretical foundations and crude in numerical methods. For instance, as a common practice, the coupling is one-way and implemented by linear interpolation [24,46,48]. The one-way coupling, i.e., from a far field to its near field, ignores the feedback from the other direction. While it captures physical phenomena in limited situations, the coupling may introduce substantial uncertainties and errors, e.g., during storms in which the flows are complex and highly transient. Linear interpolation is 2nd-order accurate locally, while frequently flow solvers, e.g., FVCOM, adopt second-order accurate schemes, and thus their solutions are 3rd-order accurate locally. As a result, the accuracy of the numerical solution degenerates at the interfaces. In the case of two-way coupling, the coupling issue becomes more complicated and challenging to study. Additional issues are stability, convergence, acceleration of computation, etc. In the past, only minimal efforts have been made on analysis on these issues, e.g., interface conditions, algorithms, optimal Schwarz iteration [61,62,64]. These issues present a great challenge to us in practical computations. Examples include numerical oscillations, delay of response in solution, and even non-physical solutions occurring at interfaces [64]. Domain decomposition is a broad framework for coupling of different models, and its research has been extensive in the mathematics community, e.g., [12,14], but with very sparse efforts for problems of interests to the ocean science community, e.g., references [27,28] collected in this Special Issue. Therefore, advances of domain decomposition methods are yet to be made directly for ocean flows. Sufficient research on all of these issues with rigorous foundations and better ways is indispensable before multiscale and multiphysics simulations become widespread within the oceanography community and reach the capability levels of directly dealing with real-world problems like the example problems illustrated in Section 1.

Another indispensable aspect of future efforts is the measurement of data in laboratory experiments and field observations. Due to various uncertainties and shortages in models, measured data play a crucial role in model validation and calibration, assuring they work correctly and reliably. Besides regional data primarily associated with conventional large-scale modeling approaches, measurements that reflect multiscale and multiphysics features, particularly those for small-scale, local flows, are highly desirable. For example, in an attempt to model correlations between background ocean currents and mixing flows generated by an offshore floating windmill farm, simultaneous observations in the far fields as well as in the near fields of each floating device are needed. In some situations, e.g., storm surge impinges coastal infrastructures, special data for engineering purposes, such as those on impinging pressure on the infrastructure, are desired, e.g., in simulations of the impact loads. In the past, numerous measurement programs have focused on revealing large-scale phenomena and their interaction, such as ocean tides, estuarine circulation, and offshore upwelling [65,66]. However, there is lack of measurements for local phenomena, especially those in conjunction with those for background flows, and data archives that characterize multiscale, multiphysics features are minimal. For instance, although an intensive experimental study has been made to understand the impact of tsunamis on coastal structures in recent years, they are focused on local flows without consideration or with substantial simplification for actual real background far-field flows, e.g., [67]. As a result, mostly the newly developed models with multiscale, multiphysics capabilities, such as the modeling system used to produce the simulation in Figure 2, have not fully be validated and tested by data with multiscale, multiphysics information. Therefore, obtaining sufficient measurement data for modeling multiscale and multiphysics processes remains a key priority and requires improved instrument networks [68,69].

As a potential direction for future development, efforts on data-driven methods and artificial intelligence are expected to grow rapidly, which could lead to new avenues to simulations of multiscale and multiphysics ocean flows. Since such flows result from various factors (e.g., wind, tide, bathymetry), their interactions, and associated uncertainties (e.g., randomness of wind), it has been a challenging task to reliably to take all of them into consideration using conventional physical and deterministic approaches as discussed above. Artificial intelligence is based on datasets that contain the info of such factors, and data-driven methods could overcome the challenges. The ideas to study ocean flows via a data-driven approach started a long time ago, such as using artificial neural networks to identify ocean currents with satellite images [70], predict storm surge with data of wind, air pressure, and tidal level [71], simulate ocean-water overtopping at structures [72], estimate waves using observed and modeled data [73], track ocean drifters according to their motion histories [74]. Now, it has been recognized in the ocean science community that data-driven artificial intelligence will be a future direction [75]. In recent years, the progress in artificial intelligence is encouraging on topics of resolving complex physics and rigorous foundations; they cover reproducing flow patterns [76], solving the Navier–Stokes equations [77], constructing turbulence closures [78], and exploring the mathematical foundation of machine learning [79]. Figure 3 shows the prediction capability of machine learning for a cavity flow. In this example, 20 images of the velocity field (vertical velocity, w) at different Reynolds numbers Re = 300, ... , 390, 410, ... , 500 (without that at Re = 400) as training data are used to train neural networks. The training uses the velocity at the upper and lower parts as input and velocity in the middle zone, an interface zone, as output. It is seen that then the trained networks satisfactorily predict the velocity at Re = 400 in the interface zone. Note that, at the same time, machine learning tools have become mature, such as Tensorflow [80], Pythoch [81], and Matlab [82], and they are open-source for application to various problems. Very recently, it is proposed to couple differential equations and flows using machine learning [83,84]. For instance, it is shown in [84] that, after being trained by solutions of partial differential equations with an initial condition, neural-network-based interface algorithms work well in solving the equations with another initial value condition. This manifests that machine learning does not just repeat its training data but also exhibits a certain prediction capability. Although it has not been realized yet, all of these indicate that machine learning could lead to avenues to simulations of multiscale and multiphysics ocean flows.

In efforts towards simulations of multiscale, multiphysics coastal ocean flows to meet the emerging needs in practice, such as the example problems described in Section 1, a multidisciplinary effort is dispensable. As described above and recognized by many researchers [85,86], collaboration among different communities such as ocean science, mathematics, and engineering lead to innovation. In addition, since such simulations cover various data, models, and applications, teamwork and collaboration across institutions are essential [14,86,87].

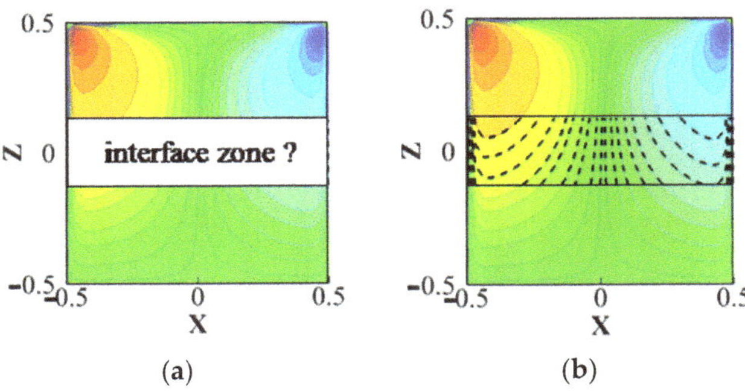

Figure 3. Example of machine learning for an unsteady, 3D cavity flow (The fluid inside a cubic box in size of $1 \times 1 \times 1$ is initially stationary and then starts to flow due to motion of the top wall at speed of cos(2πt)). Re = 400, and t = 0.25. Solid lines—data, dash lines—prediction. Tensorflow is used [80]. (**a**) Training data. (**b**) Data and prediction.

5. Concluding Remarks

Development in simulations of multiscale and multiphysics coastal ocean flows will promote research on many emerging problems resulting from changing coastal environments. This paper presents a review and discussion of such simulations, and it leads to the following conclusions:

1. Multiscale simulation has become widespread, while the multiphysics simulation remains in the preliminary stages of research
2. Model coupling is considered the most feasible and promising approach to realizing multiscale, multiphysics ocean flows for the foreseeable future, given the status of techniques and interests of funding programs.
3. Future multiscale and multiphysics research efforts will be based on rigorous foundations and methods, field data collection, and data-driven artificial intelligence.

Advances on topics such as machine learning will lead to new opportunities and breakthroughs in simulations of multiscale and multiphysics coastal oceans. Besides, the increasing needs from the communities and further development in relevant areas, e.g., computer power, will also promote the simulations. Optimistically we anticipate that understanding of multiscale and multiphysics coastal ocean phenomena will progress substantially in the coming decade.

Author Contributions: Conceptualization, H.T., C.R.N., L.D.W. and D.R.; writing—original draft preparation, H.T. and C.R.N.; writing—review and editing, H.T., C.R.N., L.D.W. and D.R. All authors have read and agreed to the published version of the manuscript.

Funding: This research received no external funding.

Institutional Review Board Statement: Not applicable.

Informed Consent Statement: Not applicable.

Data Availability Statement: No new data were created in this review article.

Acknowledgments: SURA's Coastal and Environmental Research Committee continues to promote the transdisciplinary, multi-institutional initiatives that underpin this Special Issue. The authors thank K. Qu and S. S. Li for producing some figures in this paper. They are grateful to Esme Wang and other journal editors for the help in handling and publishing the issue.

Conflicts of Interest: The authors declare no conflict of interest.

References

1. Carriger, J.F.; Jordan, S.J.; Kurtz, J.C.; Benson, W.H. Identifying evaluation considerations for the recovery and restoration from the 2010 Gulf of Mexico oil spill: An initial appraisal of stakeholder concerns and values. *Integr. Environ. Assess. Manag.* **2015**, *11*, 502–513. [CrossRef]
2. Pietrafesa, L.J.; Zhang, H.; Bao, S.; Gayes, P.T.; Hallstrom, J.O. Coastal Flooding and Inundation and Inland Flooding due to Downstream Blocking. *J. Mar. Sci. Eng.* **2019**, *7*, 336. [CrossRef]
3. Neill, S.P.; Angeloudis, A.; Robins, P.E.; Walkington, I.; Ward, S.L.; Masters, I.; Lewis, M.J.; Piano, M.; Avdis, A.; Piggott, M.D.; et al Tidal range energy resource and optimization—Past perspectives and future challenges. *Renew. Energy* **2018**, *127*, 763–778 [CrossRef]
4. Nichols, C.R.; Wright, L.D. The Evolution and Outcomes of a Collaborative Testbed for Predicting Coastal Threats. *J. Mar. Sci. Eng.* **2020**, *8*, 612. [CrossRef]
5. The Princeton Model. Available online: http://www.ccpo.odu.edu/POMWEB/ (accessed on 27 July 2021).
6. ADCIRC. Available online: https://adcirc.org/community/ (accessed on 27 July 2021).
7. Regional Ocean Modeling System (ROMS). Available online: https://www.myroms.org/ (accessed on 30 July 2021).
8. Wave Watch III. Available online: https://polar.ncep.noaa.gov/waves/wavewatch/ (accessed on 15 March 2021).
9. Qu, K. Computational Study of Hydrodynamic Impact by Extreme Surge and Wave on Coastal Structure. Ph.D. Thesis, The City College of New York, New York, NY, USA, 2007.
10. Lermusiaux, P.F.J.; Schröter, J.; Danilov, S.; Iskandarani, M.; Pinardi, N.; Westerink, J.J. Multiscale modeling of coastal, shelf, and global ocean dynamics. *Ocean Dyn.* **2013**, *63*, 1341–1344. [CrossRef]
11. Haidvogel, D.B.; Curchitser, E.N.; Danilov, S.; Fox-Kemper, B. Numerical modelling in a multiscale ocean. *J. Mar. Res.* **2017**, *75*, 683–725. [CrossRef]
12. Tang, H.S.; Haynes, R.D.; Houzeaux, G. A Review of Domain Decomposition Methods for Simulation of Fluid Flows: Concepts, Algorithms, and Applications. *Arch. Comput. Methods Eng.* **2021**, *28*, 841–873. [CrossRef]
13. Kondic, L.; González, A.G.; Diez, J.A.; Fowlkes, J.D.; Rack, P. Liquid-State Dewetting of Pulsed-Laser-Heated Nanoscale Metal Films and Other Geometries. *Annu. Rev. Fluid Mech.* **2020**, *52*, 235–262. [CrossRef]
14. Keyes, D.E.; McInnes, L.C.; Woodward, C.; Gropp, W.; Myra, E.; Pernice, M.; Bell, J.; Brown, J.; Clo, A.; Connors, J.; et al. Multiphysics simulations: Challenges and opportunities. *Int. J. High Perform. Comput. Appl.* **2013**, *27*, 4–83. [CrossRef]
15. Wu, X.-G.; Tang, H.-S. Coupling of CFD model and FVCOM to predict small-scale coastal flows. *J. Hydrodyn. Ser. B* **2010**, *22*, 284–289. [CrossRef]
16. Tang, H.S.; Keen, T. Hybrid model approaches to predict multiscale and multiphysics coastal hydrodynamic and sediment transport processes. In *Sediment Transport*; Ginsberg, S.S., Ed.; Intech: London, UK, 2011; ISBN 978-953-307-189-3.
17. Candy, A.S. An implicit wetting and drying approach for non-hydrostatic baroclinic flows in high aspect ratio domains. *Adv. Water Resour.* **2017**, *102*, 188–205. [CrossRef]
18. Tang, H.S.; Nichols, C.R.; Resio, D.T.; Wright, D. Multiscale, Multiphysics Modelling of Coastal Ocean Processes: Paradigms and Approaches, J. Marine Science and Engineering, Special Issue. 2019. Available online: https://www.mdpi.com/journal/jmse/special_issues/mul_model_coastal (accessed on 27 July 2021).
19. Loftis, J.D.; Mitchell, M.; Schatt, D.; Forrest, D.R.; Wang, H.V.; Mayfield, D.; Stiles, W.A. Validating an Operational Flood Forecast Model Using Citizen Science in Hampton Roads, VA, USA. *J. Mar. Sci. Eng.* **2019**, *7*, 242. [CrossRef]
20. Huang, H.; Chen, C.; Cowles, G.W.; Winant, C.D.; Beardsley, R.C.; Hedstrom, K.S.; Haidvogel, D.B. FVCOM validation experiments: Comparisons with ROMS for three idealized barotropic test problems. *J. Geophys. Res.* **2008**, *113*, c07042. [CrossRef]
21. Nudds, S.; Lu, Y.; Higginson, S.; Haigh, S.P.; Paquin, J.-P.; O'Flaherty-Sproul, M.; Taylor, S.; Blanken, H.; Marcotte, G.; Smith, G.C.; et al. Evaluation of Structured and Unstructured Models for Application in Operational Ocean Forecasting in Nearshore Waters. *J. Mar. Sci. Eng.* **2020**, *8*, 484. [CrossRef]
22. Iuppa, C.; Cavallaro, L.; Vicinanza, D.; Foti, E. Investigation of suitable sites for wave energy converters around Sicily (Italy). *Ocean. Sci.* **2015**, *11*, 543–557. [CrossRef]
23. Huang, W.; Li, C. Contrasting Hydrodynamic Responses to Atmospheric Systems with Different Scales: Impact of Cold Fronts vs. That of a Hurricane. *J. Mar. Sci. Eng.* **2020**, *8*, 979. [CrossRef]
24. Harris, C.; Syvitski, J.; Arango, H.; Meiburg, E.; Cohen, S.; Jenkins, C.; Birchler, J.; Hutton, E.; Kniskern, T.; Radhakrishnan, S.; et al. Data-Driven, Multi-Model Workflow Suggests Strong Influence from Hurricanes on the Generation of Turbidity Currents in the Gulf of Mexico. *J. Mar. Sci. Eng.* **2020**, *8*, 586. [CrossRef]
25. Wang, J.Y.; Tang, H.S.; Fang, H.W. A fully-coupled method to simulate wave, current, and morphology system. *Comm. Nonlinear Sci. Numer. Simul.* **2013**, *18*, 1694–1709. [CrossRef]
26. Oishi, Y.; Piggott, M.D.; Maeda, T.; Kramer, S.C.; Collins, G.S.; Tsushima, H.; Furumura, T.; Kramer, S. Three-dimensional tsunami propagation simulations using an unstructured mesh finite element model. *J. Geophys. Res. Solid Earth* **2013**, *118*, 2998–3018. [CrossRef]
27. Valera, M.; Thomas, M.P.; Garcia, M.; Castillo, J.E. Parallel Implementation of a PETSc-Based Framework for the General Curvilinear Coastal Ocean Model. *J. Mar. Sci. Eng.* **2019**, *7*, 185. [CrossRef]
28. Agoshkov, V.; Lezina, N.; Sheloput, T. Domain Decomposition Method for the Variational Assimilation of the Sea Level in a Model of Open Water Areas Hydrodynamics. *J. Mar. Sci. Eng.* **2019**, *7*, 195. [CrossRef]

29. Agoshkov, V.I.; Zalesny, V.B.; Sheloput, T.O. Variational Data Assimilation in Problems of Modeling Hydrophysical Fields in Open Water Areas. *Izv. Atmos. Ocean. Phys.* **2020**, *56*, 253–267. [CrossRef]

30. Ringler, T.; Petersen, M.; Higdon, R.L.; Jacobsen, D.; Jones, P.W.; Maltrud, M. A multi-resolution approach to global ocean modeling. *Ocean Model.* **2013**, *69*, 211–232. [CrossRef]

31. Wang, Q.; Wekerle, C.; Danilov, S.; Wang, X.; Jung, T. A 4.5 km resolution Arctic Ocean simulation with the global multi-resolution model FESOM 1.4. *Geosci. Model Dev.* **2018**, *11*, 1229–1255. [CrossRef]

32. Tang, H.; Kraatz, S.; Qu, K.; Chen, G.; Aboobaker, N.; Jiang, C. High-resolution survey of tidal energy towards power generation and influence of sea-level-rise: A case study at coast of New Jersey, USA. *Renew. Sustain. Energy Rev.* **2014**, *32*, 960–982. [CrossRef]

33. Ribal, A.; Babanin, A.V.; Zieger, S.; Liu, Q. A high-resolution wave energy resource assessment of Indonesia. *Renew. Energy* **2020**, *160*, 1349–1363. [CrossRef]

34. Blumberg, A.F.; Georgas, N.; Yin, L.; Herrington, T.O.; Orton, P.M. Street-Scale Modeling of Storm Surge Inundation along the New Jersey Hudson River Waterfront. *J. Atmos. Ocean. Technol.* **2015**, *32*, 1486–1497. [CrossRef]

35. Biastoch, A.; Sein, D.; Durgadoo, J.; Wang, Q.; Danilov, S. Simulating the Agulhas system in global ocean models—Nesting vs. multi-resolution unstructured meshes. *Ocean Model.* **2018**, *121*, 117–131. [CrossRef]

36. Greenberg, D.A.; Dupont, F.; Lyard, F.H.; Lynch, D.R.; Werner, F.E. Resolution issues in numerical models of oceanic and coastal circulation. *Cont. Shelf Res.* **2007**, *27*, 1317–1343. [CrossRef]

37. Fringer, O.; McWilliams, J.; Street, R. A New Hybrid Model for Coastal Simulations. *Oceanography* **2006**, *19*, 64–77. [CrossRef]

38. Qia, J.; Chen, C.C.; Beardsley, R.C.; Perrie, W.; Cowles, G.W.; Lai, Z. An unstructured-grid finite-volume surface wave model (FVCOM-SWAVE): Implementation, validations and applications. *Ocean. Model.* **2009**, *28*, 153–166. [CrossRef]

39. Bennis, A.-C.; Ardhuin, F.; Dumas, F. On the coupling of wave and three-dimensional circulation models: Choice of theoretical framework, practical implementation and adiabatic tests. *Ocean Model.* **2011**, *40*, 260–272. [CrossRef]

40. Couvelard, X.; Lemarié, F.; Samson, G.; Redelsperger, J.-L.; Ardhuin, F.; Benshila, R.; Madec, G. Development of a two-way-coupled ocean—Wave model: Assessment on a global NEMO(v3.6)–WW3(v6.02) coupled configuration. *Geosci. Model Dev.* **2020**, *13*, 3067–3090. [CrossRef]

41. Warner, J.C.; Sherwood, C.R.; Signell, R.; Harris, C.; Arango, H.G. Development of a three-dimensional, regional, coupled wave, current, and sediment-transport model. *Comput. Geosci.* **2008**, *34*, 1284–1306. [CrossRef]

42. Abolfazli, E.; Liang, J.; Fan, Y.; Chen, Q.J.; Walker, N.D.; Liu, J. Surface Gravity Waves and Their Role in Ocean-Atmosphere Coupling in the Gulf of Mexico. *J. Geophys. Res. Oceans* **2020**, *125*, 014820. [CrossRef]

43. Bai, P.; Wang, J.; Chu, P.; Hawley, N.; Fujisaki-Manome, A.; Kessler, J.; Lofgren, B.M.; Beletsky, D.; Anderson, E.J.; Li, Y. Modeling the ice-attenuated waves in the Great Lakes. *Ocean Dyn.* **2020**, *70*, 991–1003. [CrossRef]

44. Chen, X. Dynamic coupling of a three-dimensional hydrodynamic model with a laterally averaged, two-dimensional hydrodynamic model. *J. Geophys. Res.* **2007**, *112*, c07022. [CrossRef]

45. Tang, H.; Kraatz, S.; Wu, X.; Cheng, W.; Qu, K.; Polly, J. Coupling of shallow water and circulation models for prediction of multiphysics coastal flows: Method, implementation, and experiment. *Ocean Eng.* **2013**, *62*, 56–67. [CrossRef]

46. Fujima, K.; Masamura, K.; Goto, C. Development of the 2D/3D Hybrid Model for Tsunami Numerical Simulation. *Coast. Eng. J.* **2002**, *44*, 373–397. [CrossRef]

47. Choboter, P.F.; Garcia, M.; De Cecchis, D.; Thomas, M.; Walter, R.K.; Castillo, J.E. Nesting nonhydrostatic GCCOM within hydrostatic ROMS for multiscale coastal ocean modeling. In Proceedings of the OCEANS 2016 MTS/IEEE Monterey, Monterey, CA, USA, 19–23 September 2016; pp. 1–4. Available online: https://ieeexplore.ieee.org/document/7761488 (accessed on 27 July 2021).

48. Tang, H.; Qu, K.; Wu, X. An overset grid method for integration of fully 3D fluid dynamics and geophysics fluid dynamics models to simulate multiphysics coastal ocean flows. *J. Comput. Phys.* **2014**, *273*, 548–571. [CrossRef]

49. Qu, K.; Tang, H.; Agrawal, A. Integration of fully 3D fluid dynamics and geophysical fluid dynamics models for multiphysics coastal ocean flows: Simulation of local complex free-surface phenomena. *Ocean Model.* **2019**, *135*, 14–30. [CrossRef]

50. Perezhogin, P.; Chernov, I.; Iakovlev, N. Advanced parallel implementation of the coupled ocean–ice model FEMAO (version 2.0) with load balancing. *Geosci. Model Dev.* **2021**, *14*, 843–857. [CrossRef]

51. Sein, D.V.; Groger, M.; Cabos, W.; Alvarez-Garcia, F.J.; Hagemann, S.; Pinto, J.G.; Izquierdo, A.; de la Vara, A.; Koldunov, N.V.; Dvornikov, A.Y.; et al. Regionally coupled atmosphere-ocean-marine biogeochemistry model ROM: 2. Studying the climate change signal in the North Atlantic and Europe. *J. Adv. Modeling Earth Syst.* **2020**, *12*, e2019MS001646.

52. Wahl, T.; Jain, S.; Bender, J.; Meyers, S.; Luther, M.E. Increasing risk of compound flooding from storm surge and rainfall for major US cities. *Nat. Clim. Chang.* **2015**, *5*, 1093–1097. [CrossRef]

53. Allard, R.; Rogers, E.; Martin, P.; Jensen, T.; Chu, P.; Campbell, T.; Dykes, J.; Smith, T.; Choi, J.; Gravois, U. The US Navy Coupled Ocean-Wave Prediction System. *Oceanography* **2014**, *27*, 92–103. [CrossRef]

54. Qi, J.; Chen, C.; Beardsley, R.C. FVCOM one-way and two-way nesting using ESMF: Development and validation. *Ocean Model.* **2018**, *124*, 94–110. [CrossRef]

55. Brown, A.J.G.; Neill, S.P.; Lewis, M.J. Tidal energy extraction in three-dimensional ocean models. *Renew. Energy* **2017**, *114*, 244–257. [CrossRef]

56. Paulsen, B.T.; Bredmose, H.; Bingham, H.B. An efficient domain decomposition strategy for wave loads on surface piercing circular cylinders. *Coast. Eng.* **2014**, *86*, 57–76. [CrossRef]

57. Blain, C.A.; Massey, T.C. Application of a coupled discontinuous–continuous Galerkin finite element shallow water model to coastal ocean dynamics. *Ocean Model.* **2005**, *10*, 283–315. [CrossRef]

58. Botelho, D.; Imberger, J.; Dallimore, C.; Hodges, B.R. A hydrostatic/non-hydrostatic grid-switching strategy for computing high-frequency, high wave number motions embedded in geophysical flows. *Environ. Model. Softw.* **2009**, *24*, 473–488. [CrossRef]

59. Hoang, T.-T.-P.; Leng, W.; Ju, L.; Wang, Z.; Pieper, K. Conservative explicit local time-stepping schemes for the shallow water equations. *J. Comput. Phys.* **2019**, *382*, 152–176. [CrossRef]

60. Debreu, L.; Marchesiello, P.; Penven, P.; Cambon, G. Two-way nesting in split-explicit ocean models: Algorithms, implementation, and validation. *Ocean Model.* **2012**, *49*, 1–21. [CrossRef]

61. Blayo, E.; Rousseau, A. About interface conditions for coupling hydrostatic and nonhydrostatic Navier-Stokes flows. *Discret. Contin. Dyn. Syst. Ser. S* **2016**, *9*, 1565–1574. [CrossRef]

62. Tang, H.; Liu, Y. Coupling of Navier-Stokes Equations and Their Hydrostatic Versions for Ocean Flows: A Discussion on Algorithm and Implementation. In *Domain Decomposition Methods in Science and Engineering XXV*; DD 2018. Lecture Notes in Computational Science and Engineering; Haynes, R., MacLachlan, S., Cai, X.-C., Halpern, L., Kim, H.H., Klawonn, A., Widlund O., Eds.; Springer: Berlin, Germany, 2020; p. 138.

63. Connors, J.M.; Dolan, R.D. Stability of two conservative, high-order fluid-fluid coupling methods. *Adv. Appl. Math. Mech.* **2019**, *11*, 1287–1338.

64. Tang, H.S.; Qu, K.; Wu, X.G.; Zhang, Z.K. Domain Decomposition for a Hybrid Fully 3D Fluid Dynamics and Geophysical Fluid Dynamics Modeling System: A Numerical Experiment on Transient Sill Flow. In *Mesh free Methods for Partial Differential Equations VIII*; Springer Science and Business Media: Berlin, Germany, 2016; pp. 407–414.

65. NOAA NDBC. 2021. Available online: https://www.ndbc.noaa.gov/ (accessed on 27 July 2021).

66. Talley, L.; Pickard, G.L.; Emery, W.J.; Swift, J.H. *Descriptive Physical Oceanography: An introduction*, 6th ed.; Elsevier: Boston, MA, USA, 2011.

67. Chen, X.; Hofland, B.; Altomare, C.; Suzuki, T.; Uijttewaal, W. Forces on a vertical wall on a dike crest due to over-topping flow. *Coast. Eng.* **2015**, *95*, 94–104. [CrossRef]

68. Subramanian, A.C.; Balmaseda, M.A.; Centurioni, L.; Chattopadhyay, R.; Cornuelle, B.D.; DeMott, C.; Flatau, M.; Fujii, Y.; Giglio, D.; Gille, S.T.; et al. Ocean Observations to Improve Our Understanding, Modeling, and Forecasting of Subseasonal-to-Seasonal Variability. *Front. Mar. Sci.* **2019**, *6*, 427. [CrossRef]

69. Nichols, C.R.; Raghukumar, K. *Marine Environmental Characterization*; Morgan & Claypool Publishers: San Rafael, CA, USA, 2020. [CrossRef]

70. Cote, S.; Tatnall, A.R.L. Estimation of ocean surface currents from satellite imagery using a Hopfield neural network. *Mar. Technol. Soc. J.* **1996**, *30*, 4–13.

71. Lee, T.-L. Neural network prediction of a storm surge. *Ocean Eng.* **2006**, *33*, 483–494. [CrossRef]

72. Geeraerts, J.; Troch, P.; De Rouck, J.; Verhaeghe, H.; Bouma, J. Wave overtopping at coastal structures: Prediction tools and related hazard analysis. *J. Clean. Prod.* **2007**, *15*, 1514–1521. [CrossRef]

73. James, S.C.; Zhang, Y.; O'Donncha, F. A machine learning framework to forecast wave conditions. *Coast. Eng.* **2018**, *137*, 1–10. [CrossRef]

74. Aksamit, N.O.; Sapsis, T.; Haller, G. Machine-Learning Mesoscale and Submesoscale Surface Dynamics from Lagrangian Ocean Drifter Trajectories. *J. Phys. Oceanogr.* **2020**, *50*, 1179–1196. [CrossRef]

75. Fringer, O.B.; Dawson, C.N.; He, R.; Ralston, D.K.; Zhang, Y.J. The future of coastal and estuarine modeling: Findings from a workshop. *Ocean Model.* **2019**, *143*, 101458. [CrossRef]

76. Wiewel, S.; Becher, M.; Thuerey, N. Latent Space Physics: Towards Learning the Temporal Evolution of Fluid Flow. *Comput. Graph. Forum* **2019**, *38*, 71–82. [CrossRef]

77. Baymani, M.; Effati, S.; Niazmand, H.; Kerayechian, A. Artificial neural network method for solving the Navier–Stokes equations. *Neural Comput. Appl.* **2014**, *26*, 765–773. [CrossRef]

78. San, O.; Maulik, R. Extreme learning machine for reduced order modeling of turbulent geophysical flows. *Phys. Rev. E* **2018**, *97*, 042322. [CrossRef]

79. Sirignan, J.; Spiliopoulos, K. DGM: A deep learning algorithm for solving partial differential equations. *J. Comput. Phys.* **2018**, *375*, 1339–1364. [CrossRef]

80. Tensorflow, An End-to-End Open Source Machine Learning Platform. Available online: https://www.tensorflow.org/ (accessed on 27 July 2021).

81. PyTorch. From Research to Production. Available online: https://pytorch.org/ (accessed on 2 August 2021).

82. MathWorks, "MATLAB". Available online: https://www.mathworks.com/products/matlab.html (accessed on 82 July 2021).

83. Pawar, S.; Ahmed, S.E.; San, O. Interface learning in fluid dynamics: Statistical inference of closures within micro–macro-coupling models featured. *Phys. Fluids* **2020**, *32*, 091704. [CrossRef]

84. Tang, H.; Li, L.; Grossberg, M.; Liu, Y.; Jia, Y.; Li, S.; Dong, W. An exploratory study on machine learning to couple numerical solutions of partial differential equations. *Commun. Nonlinear Sci. Numer. Simul.* **2021**, *97*, 105729. [CrossRef]

85. Wright, L.D.; Nichols, C.R. (Eds.) *Tomorrow's Coasts: Complex and Impermanent*; Coastal Research Library; Springer: Cham, Switzerland, 2019; Volume 27.

86. Levine, R.M.; Fogaren, K.E.; Rudzin, J.E.; Russoniello, C.J.; Soule, D.C.; Whitaker, J.M. Open data, collaborative working platforms, and interdisciplinary collaboration: Building an early career scientist community of practice to lever-age ocean observatories initiative data to address critical questions in marine science. *Front. Mar. Sci.* **2020**, *7*, 1011. [CrossRef]
87. Deng, Z.; Xie, L.; Liu, B.; Wu, K.; Zhao, D.; Yu, T. Coupling winds to ocean surface currents over the global ocean. *Ocean Model.* **2009**, *29*, 261–268. [CrossRef]

MDPI

St. Alban-Anlage 66

4052 Basel

Switzerland

Tel. +41 61 683 77 34

Fax +41 61 302 89 18

www.mdpi.com

Journal of Marine Science and Engineering Editorial Office

E-mail: jmse@mdpi.com

www.mdpi.com/journal/jmse